SpringerBriefs in Applied Sciences and Technology

Computational Mechanics

Series Editors

Holm Altenbach, Faculty of Mechanical Engineering,
Otto-von-Guericke-Universität Magdeburg, Magdeburg, Sachsen-Anhalt, Germany

Lucas F. M. da Silva, Department of Mechanical Engineering, Faculty of
Engineering, University of Porto, Porto, Portugal

Andreas Öchsner, Faculty of Mechanical Engineering, Esslingen University of
Applied Sciences, Esslingen, Germany

These SpringerBriefs publish concise summaries of cutting-edge research and practical applications on any subject of computational fluid dynamics, computational solid and structural mechanics, as well as multiphysics.

SpringerBriefs in Computational Mechanics are devoted to the publication of fundamentals and applications within the different classical engineering disciplines as well as in interdisciplinary fields that recently emerged between these areas.

More information about this subseries at http://www.springer.com/series/8886

Alberto Pozo Álvarez

Fluid Mechanics Applied to Medicine

Cardiac Flow Visualization Techniques

 Springer

Alberto Pozo Álvarez ⓘ
Department of Industrial Engineering
University of Valladolid
Valladolid, Spain

ISSN 2191-530X   ISSN 2191-5318   (electronic)
SpringerBriefs in Applied Sciences and Technology
ISSN 2191-5342   ISSN 2191-5350   (electronic)
SpringerBriefs in Computational Mechanics
ISBN 978-3-030-60388-5   ISBN 978-3-030-60389-2   (eBook)
https://doi.org/10.1007/978-3-030-60389-2

This Springer imprint is published by the registered company Springer Nature Switzerland AG
The registered company address is: Gewerbestrasse 11, 6330 Cham, Switzerland

The original version of the book was revised: The author's last name in Ref. 3, 4, 9 of Chap. 3, Ref. 22, 23, 30, 33 of Chap. 4 and Ref. 2, 3, 6, 8, 13, 15 of Chap. 5 were incorrect. The author's last names are corrected in References. The correction to the book is available at https://doi.org/10.1007/978-3-030-60389-2_6

Contents

Symbols and Acronyms

LAA	Left Atrial Appendage
CFD	Computational Fluid Dynamics
ECG	Electrocardiogram
PIV	Particle Image Velocimetry
Echo-PIV	Echocardiographic Particle Image Velocimetry
Micro-PIV	Microscopic Particle Image Velocimetry
PTV	Particle Tracking Velocimetry
LDV	Laser Doppler Velocimetry
MRI	Magnetic Resonance Imaging
PC-MRI	Phase Contrast Magnetic Resonance Imaging
PC-MRA	Phase Contrast Magnetic Resonance Angiography
TEE	Transesophageal Echocardiography
TTE	Transthoracic Echocardiography
ICE	Intracardiac Echocardiography
CT	Computed Tomography
MDCT	Multiple Detector Computed Tomography
PET	Positron Emission Tomography
SPECT	Single-Photon Emission Computed Tomography
SNR	Signal-to-Noise Ratio
FOV	Field of View
B	Magnetic field
α	Ernst angle
Re	Reynolds number
Wo	Womersley number
P	Pressure
Q	Flow rate
e	Internal energy per mass unit

ρ	Density
μ	Dynamic viscosity
g	Gravitational acceleration
T	Temperature

List of Figures

List of Tables

Chapter 1
Introduction

1.1 Justification

Heart disease is the first cause of global mortality, as it can be seen in Fig. 1.1, whose data have been extracted from the World Health Organization (WHO).

Ischemic heart disease is the disease with the highest mortality rate. Within this group, the main disease is **atherosclerosis**, consisting of a localized narrowing in arteries caused by the accumulation of atheroma plaque, which hinders the blood flow supplied to the heart. This reduction of the section in arteries can cause angina pectoris if it is partial or a heart attack if the obstruction is complete and sudden.

The most widely used method to combat atherosclerotic stenosis is the stent, a tubular metal mesh that is implanted inside the affected artery to recover the cross-sectional area, as it is shown in Fig. 1.2.

On the other hand, the second leading cause of mortality worldwide is cerebrovascular diseases such as hemorrhage, effusion, embolism, thrombosis, cerebral stroke or stroke.

A **embolism** is the obstruction of a blood vessel by a mass, which does not dissolve in the blood, an embolus that has been carried away by the bloodstream. When the embolism is caused by a thrombus (blood clot), it is called **thrombosis**.

A **stroke** is a sudden attack consisting of an interruption of the blood supply to the brain due to blockage of an artery (ischemic stroke) or because bleeding has occurred (hemorrhagic stroke). **The risk of stroke in patients with atrial fibrillation is 5–7 times higher** in comparison with a person who does not suffer from this arrhythmia [1].

In atrial fibrillation, the upper chambers of the heart present chaotic electrical signals causing agitation, especially in the atria. The result is a fast and irregular heartbeat. This disease is one of the most common and important cardiac affections. It is a growing problem in today's society, especially in the oldest, because from 0.4 to 1% of the world population suffers from it, increasing to more than 8% for those over 80 years [2].

A. Pozo Álvarez, *Fluid Mechanics Applied to Medicine*,
SpringerBriefs in Computational Mechanics,
https://doi.org/10.1007/978-3-030-60389-2_1

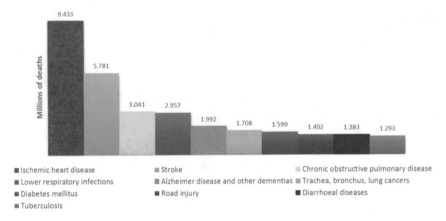

Fig. 1.1 Main causes of death in the World in 2016

Particular attention should be paid to the LAA (Left Atrial Appendage) in patients with atrial fibrillation to determine the risk of cardioembolic complications. Approximately, 90% of atrial thrombi in nonvalvular atrial fibrillation and 60% of thrombi in patients with rheumatic mitral valve disease (predominantly stenosis) are originated in the LAA [3].

The LAA is an appendix located in the Left Atrium of the heart. Atrial fibrillation is thought to produce recirculations of blood in the atrium that may promote thrombus formation in the LAA, where flow conditions are closer to stagnation. In a study of Biase et al. [4], the LAA appears to be responsible for cardiac arrhythmias in 27% of patients (Fig. 1.3).

In recent years, it has been demonstrated the usefulness of CFD (Computational Fluid Mechanics) supported by new imaging or diagnostic techniques for the study of cardiovascular diseases, treatment improvement, and prevention [5]. These techniques allow a personalized study of the patient, through the digital twin or virtual patient. These *ad hoc* studies are considered as one of the challenges with the greatest potential impact on cardiovascular biomechanics [6].

One of the main problems that arise within the analysis of virtual patients is the validation of computational codes and their reproducibility [7]. Being numerical models, their results must be validated with experimental tests that confirm their behavior. Of the existing types of validation, two are mainly used: in vitro validation [8] and in vivo validation [9].

In vivo techniques are those that are carried out directly on the patient's own organism, so in principle they better reflect reality than in vitro techniques, which are carried out outside the living organism. As it is not always possible to perform in vivo tests, with in vitro techniques a much more controlled environment is achieved, which makes them ideal for code validation.

In vitro techniques are based on the use of a fluid with characteristics similar to blood flow that, at the same time, is transparent and with a refractive index similar

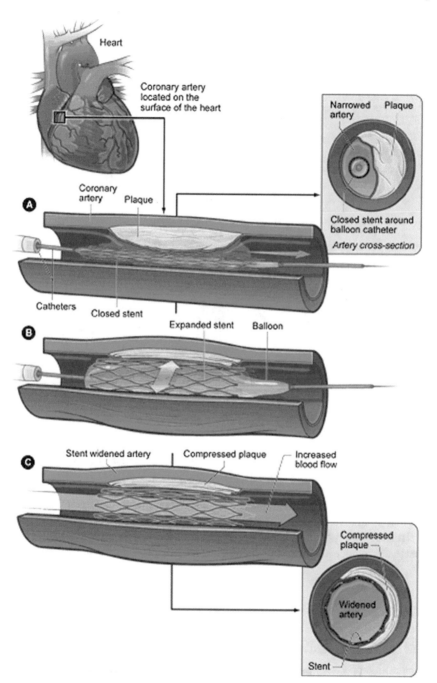

Fig. 1.2 Stent placement using a catheter (Creative Commons License (CC0): public domain)

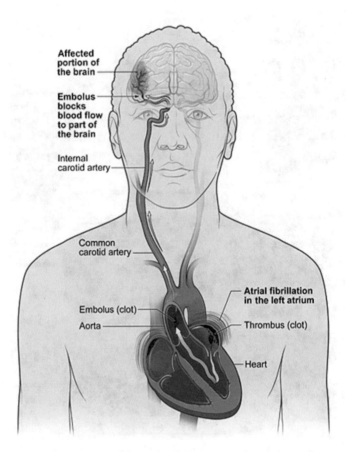

Fig. 1.3 Stroke as a consequence of the formation of a thrombus in the Left Atrium (Creative Commons License (CC0): public domain)

to that of the model under study in order to carry out the measurement using PIV (Particle Image Velocimetry) technique.

Treating blood as fluid is another important aspect. It is usually modeled as a Newtonian fluid, even though its rheological behavior is fundamentally different.

1.2   Content

This book is composed of the following chapters:

- **Chapter 1: Introduction**. It constitutes a presentation of the book from an overview. It is justified the use of CFD for the study of cardiovascular diseases.

- **Chapter 2: Fluid mechanics history**. It constitutes an approximation to Fluid Mechanics from the historical point of view. Special attention will be paid to the Navier–Stokes Equations, which govern the movement of fluids.
- **Chapter 3: Fluid-mechanical description of blood flow**. This is an analysis from the engineering point of view of the characteristics of blood flow, covering theoretical and experimental fundamentals. In this chapter, a fluid dynamic characterization of the flow in the heart and blood vessels will be performed. Lastly, a description of the mathematical models of hemodynamics is presented.
- **Chapter 4: Cardiac flow visualization techniques**. Existing cardiac flow visualization techniques will be described, both in vivo and in vitro. At the end of the chapter, it will be shown the procedure to follow to validate a numerical CFD model (*in silico* technique).
- **Chapter 5: Techniques for the validation of numerical models**. The methodology to measure with the most used techniques for the validation of hemodynamic numerical models will be detailed: PC-MRI (Phase Contrast Magnetic Resonance Imaging) in case of in vivo validations and PIV for in vitro validations. A procedure to estimate measurement uncertainty is also included.

References

1. Soldevila JG, Ruíz MDM, Robert ID, Tornos P, Martínez-Rubio A (2013) Evaluación de riesgo tromboembólico y hemorrágico de los pacientes con fibrilación auricular. Revista Española de Cardiología Suplementos 13:9–13
2. Davis RC, Hobbs FDR, Kenkre JE, Roalfe AK, Iles R, Lip GYH, Davies MK (2012) Prevalence of atrial fibrillation in the general population and in high-risk groups: the ECHOES study. Europace 14(11):1553–1559
3. Blackshear JL, Odell JA (1996) Appendage obliteration to reduce stroke in cardiac surgical patients with atrial fibrillation. Ann Thoracic Surg 61(2):755–759
4. Biase LD, Burkhardt JD, Mohanty P, Sanchez J, Mohanty S, Horton R, Gallinghouse GJ, Bailey SM, Zagrodzky JD, Santangeli P, Hao S, Hongo R, Beheiry S, Themistoclakis S, Bonso A, Rossillo A, Corrado A, Raviele A, Al-Ahmad A, Wang P, Cummings JE, Schweikert RA, Pelargonio G, Russo AD, Casella M, Santarelli P, Lewis WR, Natale A (2010) Left atrial appendage. Circulation 122(2):109–118
5. Thondapu V, Bourantas CV, Foin N, Jang I-K, Serruys PW, Barlis P (2016) Biomechanical stress in coronary atherosclerosis: emerging insights from computational modelling. Eur Heart J, ehv689
6. Holmes JW, Wagenseil JE (2016) Special issue: spotlight on the future of cardiovascular engineering: Frontiers and challenges in cardiovascular biomechanics. J Biomech Eng 138(11):110301
7. Chung B, Cebral JR (2014) CFD for evaluation and treatment planning of aneurysms: review of proposed clinical uses and their challenges. Ann Biomed Eng 43(1):122–138
8. Buchmann NA, Yamamoto M, Jermy M, David T (2010) Particle image velocimetry (PIV) and computational fluid dynamics (CFD) modelling of carotid artery haemodynamics under steady flow: A validation study. J Biomech Sci Eng 5(4):421–436
9. Rayz VL, Boussel L, Acevedo-Bolton G, Martin AJ, Young WL, Lawton MT, Higashida R, Saloner D (2008) Numerical simulations of flow in cerebral aneurysms: comparison of CFD results and in vivo MRI measurements. J Biomech Eng 130(5):051011

Chapter 2
Fluid Mechanics History

2.1 Introduction

The movement of a fluid is mathematically described by the Navier–Stokes equations, which are named after Claude-Louis Navier (1785–1836) and George Gabriel Stokes (1819–1903).

It is a set of second-order nonlinear partial differential equations that describe the motion of a fluid. These equations govern the Earth's atmosphere, ocean currents, the flow around vehicles, planes, or wind turbines, and even the flow of blood in our bodies (Fig. 2.1).

Following Newtonian mechanics, these equations should determine the future motion of the fluid from its initial state. However, and despite all the efforts that have been made for more than a century, until now it has not been possible to demonstrate this determinism mathematically, nor to deny it.

In 2000, the Clay Institute of Mathematics located in Cambridge, Massachusetts, announced the seven mathematical problems of the millennium. The prize for the person who was able to solve one of them would be one million dollars. Currently, among these problems, only the Poincaré Conjecture has been solved by the Russian mathematician Grigori Perelman.

However, another of the millennium problems, the existence and smoothness of the Navier–Stokes Equations has not yet been proven.

This is because the solutions of the Navier–Stokes Equations often include the turbulence phenomenon, whose general solution is one of the biggest unsolved problems in Fluid Mechanics. Even basic properties of the solutions of the Navier–Stokes Equations have never been tested. For the 3D system of equations and given initial conditions, mathematicians have not yet shown that smooth solutions always exist, or that if they exist, they have a limited energy per mass unit. So the problem is showing that there are smooth, globally defined solutions that meet certain conditions, or that they do not always exist and the equations break down.

© The Editor(s) (if applicable) and The Author(s), under exclusive license
to Springer Nature Switzerland AG 2021
A. Pozo Álvarez, *Fluid Mechanics Applied to Medicine*,
SpringerBriefs in Computational Mechanics,
https://doi.org/10.1007/978-3-030-60389-2_2

(a) Tornado   (b) Wind turbines on a beach

Fig. 2.1 Phenomena governed by Fluid Mechanics (Creative Commons License (CC0): public domain)

As previously discussed, the Navier–Stokes equations are a set of nonlinear partial differential equations. No general solution is available for this set of equations, and except for certain types of flow and very specific situations it is not possible to find an analytical solution. That is why on numerous occasions it is necessary to use numerical analysis to determine an approximate solution. CFD is the branch of Fluid Mechanics in charge of obtaining these numerical solutions.

2.2   Navier–Stokes Equations Background

It is necessary to have a brief idea about the History of Fluid Mechanics to understand the importance of the Navier–Stokes Equations.

Fluid Mechanics is the branch of physics and mechanics, which studies the dynamics and kinematics of fluids under the action of applied forces. Within Fluid Mechanics there are subdivisions according to the type of fluid, Gas Dynamics for gases or Hydromechanics, Hydraulics, and Hydrostatics for liquids.

The History of Fluid Mechanics is parallel to the history of civilization due to the great importance that some fluids have in the development of human life, such as water. One of the great civilizations that developed water supplies and large drainage systems was the Romans, although these works were more political and artistic than scientific.

However, since the fall of the Roman Empire, advances in Fluid Mechanics have been halted in what constitutes a period of scientific lull in general.

So you can see a gap in Table 2.1 from Archimedes to Leonardo da Vinci.

Archimedes (287 BC–212 BC)
Archimedes' principle indicates that the upward thrust exerted on a body immersed in a fluid is equivalent to the weight of the fluid that the body displaces. This principle originates three different cases:

Table 2.1 Relevant discoveries throughout the History of Fluid Mechanics [1]

Author	Contribution
Archimedes (287 BC–212 BC)	Floating laws
Leonardo da Vinci (1452–1519)	Continuity equation
	Sketches of hydraulic and flying machines
Galileo Galilei (1564–1642)	Hydrostatic fundamentals
Torricelli (1608–1647)	Water outlet through an orifice
	Atmospheric pressure measurement
Pascal (1623–1662)	Pascal's law
Newton (1642–1726)	Newtonian law of viscosity
Bernoulli (1700–1782)	Bernoulli's principle
Euler (1707–1783)	Differential equations of ideal flow motion
D'Alembert (1717–1783)	Differential continuity equation
Chézy (1718–1798)	Water circulation in canals and pipes
Lagrange (1736–1813)	Potential function and streamlines
Venturi (1746–1822)	Liquid outlet through holes and nozzles
Navier (1785–1836) and Stokes (1819–1903)	Differential equations of viscous fluids motion
Poiseuille (1797–1869)	Capillary resistance equation
Darcy (1803–1858)	Pressure movement in pipes
Weisbach (1806–1871)	Pipes resistance formula
Reynolds (1842–1912)	Laminar and turbulent flow regimes
	Reynolds number
Prandtl (1868–1945)	Boundary layer theory
Blasius (1883–1970)	Laminar boundary layer solution
Von Karman (1881–1963)	Turbulent boundary layer solution

1. If the buoyant force is smaller than the weight of the body, it ends up sinking and depositing at the bottom.
2. If the buoyant force is equal to the weight of the body, it will remain submerged in the fluid but floating inside it.
3. If the buoyant force is greater than the weight of the body, it will end up ascending, keeping one part submerged and the other part protruding from the fluid.

Leonardo da Vinci (1452–1519)

Leonardo da Vinci made notable contributions to different areas of science, but especially to hydraulics. At first, he was interested in the flows that flow through the bodies, studying the trajectories of the movement of the fluid as a solid with respect to an axis, which today is called vertex movement. Other experiments he carried out led him to discover the principle of continuity: the speed of a flow varies inversely proportional to its cross-sectional area. In addition, he was the first to carry out a scientific study on the circulation of air around the Earth, that is, the birth of meteorology (Fig. 2.2).

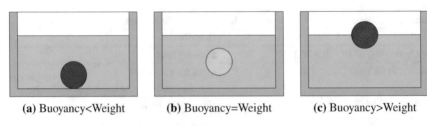

(a) Buoyancy<Weight   (b) Buoyancy=Weight   (c) Buoyancy>Weight

Fig. 2.2 Cases according to Archimedes' Principle

Fig. 2.3 Sketches of da
Vinci's flying machines
(Creative Commons License
(CC0): public domain)

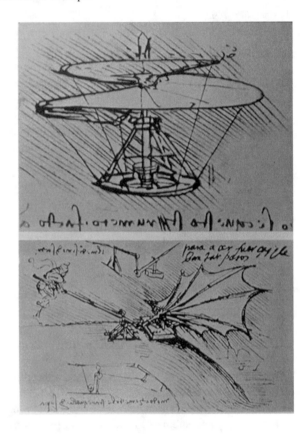

It is also credited with the design of flying machines, such as the ornithopter, which is the predecessor of the helicopter (Fig. 2.3).

Galileo Galilei (1564–1642)

He invented the hydrostatic balance, designed to study the upward thrust exerted by fluids on bodies immersed in them. It is based on Archimedes' principle and was designed to determine the densities of solids and liquids experimentally (Fig. 2.4).

It is famous his phrase, which illustrates the difficulty of Fluid Mechanics:

Fig. 2.4 Hydrostatic balance
(Creative Commons License
(CC0): public domain)

It has been easier for me to find the laws with which the heavenly bodies move, those that are millions of kilometers away, than to define the laws of the movement of water, which runs before my eyes.

Evangelista Torricelli (1608–1647)

Evangelista Torricelli formulated the equation for the average speed of a jet. He used the mercury by pushing it up into a closed tube, creating a vacuum at the top, pushed by the weight of the air in the atmosphere. He proved that air weighs, and invented the barometer. The pressure unit *torr* is named after him.

Blaise Pascal (1623–1662)

He was interested in Torricelli's work on atmospheric pressure, clarifying the principles of the barometer and distribution of pressures, thus establishing the named Pascal's Law: the pressure at a point for a fluid at rest or in motion is independent of the direction, as long as there are no tangential stresses.

Isaac Newton (1642–1727)

This English scientist who established the Universal Gravitation Law had great contributions to Fluid Mechanics. He carried out several experiments on the resistance that movement present in bodies is due to the viscosity, inertia, and elasticity of fluids. In addition, he discovered the contraction of the jets in free discharge.

Newton contributed the Law of Dynamic Viscosity, which states that the dynamic viscosity μ is the property that relates the shear stress τ in a moving fluid with the deformation speed $\delta\theta/\delta t$ of the fluid particles. The units in the IS are $kg/(m \cdot s)$. The strain rate is the relative displacement of one fluid layer with respect to the next. Intuitively, dynamic viscosity can be used to measure the resistance of a fluid to movement. Graphically, it corresponds to the value of the tangent slope at each point to the fluid velocity curve. With the hypothesis of infinitesimal deformations, the tangential deformation velocity is equivalent to the velocity gradient $\delta u/\delta y$.

Fig. 2.5 Kinds of fluids
according to its viscosity

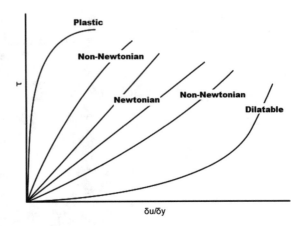

As seen in Fig. 2.5, Newtonian or homogeneous fluids are those that show a constant viscosity like water. In this case the relationship is linear, as shown by Eq. 2.1.

$$\tau = \mu \cdot \frac{\delta u}{\delta y} \tag{2.1}$$

In case of non-Newtonian or heterogeneous fluids, this dependence is not linear and the viscosity takes the value of the tangent of the curve at each point, and is no constant.

Daniel Bernoulli (1700–1782)

He experimented and wrote about various phases of fluid motion in the first known book about Fluid Mechanics, which he titled *Hydrodynamics*. His father Johan also published a book about *Hydraulics*. He invented the manometer technology and exposed what would be called the Bernoulli equation: in an ideal incompressible fluid (constant density without viscosity or friction) in circulation through a closed conduit, the energy that the fluid possesses remains constant throughout its trajectory.

$$\frac{\rho v^2}{2} + P + \rho g z = \text{constant} \tag{2.2}$$

The form of the Eq. 2.2 is due to Euler, it does not appear explicitly in the Bernoulli books. P is the pressure along the streamline, v the velocity of the fluid in the section under consideration, ρ is the density of the fluid, the acceleration due to gravity, and z the height in the direction of gravity from a reference.

Leonhard Euler (1707–1783)

Leonhard Euler is considered another of the great contributors to Fluid Mechanics with the mathematical model of fluid dynamics for ideal fluids. However, he did not consider the effect of viscosity in these mathematical models.

Euler showed how Newton's Second Law (originally applicable only to point masses) could be used to determine the acceleration of any infinitesimal part of the fluid. This fact was possible thanks to Euler's discovery of the modern concept of pressure, which led him to calculate the resultant of the contact forces, exerted by the rest of the fluid considered as a continuous medium, on this elemental mass. In 1755, he discovered the laws that govern the movement of ideal fluids, long before the discovery of the laws that govern the movement of the rigid solid. Although Euler deduced these equations directly from Newton's laws, they are usually expressed in the form of conservation or equilibrium because it is the most convenient for the computational simulation of fluid dynamics [2]. The first is the conservation of mass equation, the second is the conservation of momentum, and the third is the conservation of energy.

$$\frac{\partial \rho}{\partial t} + \nabla \cdot (\rho \mathbf{u}) \tag{2.3}$$

$$\frac{\partial \rho \mathbf{u}}{\partial t} + (\mathbf{u}\nabla) \cdot (\rho \mathbf{u}) + \nabla P = 0 \tag{2.4}$$

$$\frac{\partial E}{\partial t} + \nabla \cdot (\mathbf{u}(E + P)) = 0 \tag{2.5}$$

E is the total energy per volume unit and is calculated with the following expression:

$$E = \rho e + \frac{\rho(u^2 + v^2 + w^2)}{2} \tag{2.6}$$

In the previous expressions, e is the internal energy per mass unit for the fluid, P is the pressure, u, v, and w are the three components of the fluid velocity vector $\mathbf{u}$, and ρ the density of the fluid.

He also contributed with the Euler equation, fundamental in the field of turbomachinery. Eq. 2.7 relates the useful height transmitted by the impeller to the fluid, H_u, with the variation of the kinetic moment that the fluid experiences when passing through the impeller of a centrifugal machine under the 1D flow approximation.

$$H_u = \frac{u_2 \cdot v_{2u} - u_1 \cdot v_{1u}}{g} \tag{2.7}$$

In Eq. 2.7, g is the gravitational acceleration, u is the relative speed of the fluid, and v_u indicates the tangential absolute speed of the fluid. Subscripts 1 and 2 indicate entry and exit, respectively.

Jean le Rond D'Alembert (1717–1783)

D'Alembert introduced, at the same time as Euler (who already knew his work) a theory for the description of fluid movements, considered as continuous means. Finally, he limited himself to describing the 2D stationary movements of liquids,

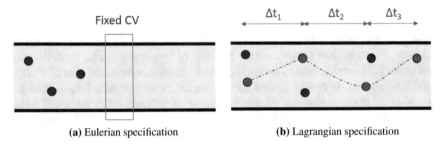

(a) Eulerian specification   **(b)** Lagrangian specification

Fig. 2.6 Types of reference systems in fluid flow

characterized by two nonzero velocity components. Obtained the continuity equation differentially:

$$\frac{\partial u}{\partial x} + \frac{\partial v}{\partial y} = 0 \tag{2.8}$$

D'Alembert is also the author of the famous paradox that bears his name, which refers to the discrepancy he found between the force of a flow of an ideal fluid (air) on a cylinder and what he observed in experiments. A century later, this paradox was clarified as there was actually a drag force on it.

Joseph-Louis Lagrange (1736–1813)

Lagrange was another of the great talents with contributions to Fluid Mechanics.

The equation known today as Bernoulli's is actually Lagrange's integral of the momentum equation presented by Euler for a fluid without viscosity.

Until then, the Eulerian specification of the fluid field dominated in Fluid Mechanics, in which the movement of the fluid was observed, focusing on fixed places in the space through which it flows as time passes. This can be visualized as sitting on the bank of a river and watching the flow from a fixed location (see Fig. 2.6a).

Instead, the Lagrangian specification of the fluid field consists of looking at the movement of the fluid so that the observer follows an individual fluid particle, indicated in red in Fig. 2.6b, while it moves through space and time. If the position of this fluid particle is traced over time, what is known as streamline (dashed red line in Fig. 2.6b) is obtained. This vision is equivalent to sitting in a boat and drifting down a river.

Relating both visions, we have the substantial or material derivative. The acceleration of a fluid particle is the substantial derivative of velocity, which is denoted as $\frac{D\mathbf{u}}{Dt}$:

$$\frac{D\mathbf{u}}{Dt} = \frac{\partial \mathbf{u}}{\partial t} + (\mathbf{u} \cdot \nabla \mathbf{u}) \tag{2.9}$$

Fig. 2.7 Venturi effect
(Creative Commons License
(CC0): public domain)

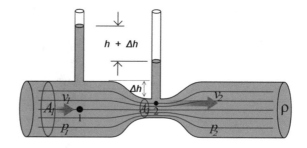

Giovanni Battista Venturi (1746–1822)

He is famous for the Venturi effect, which consists of a phenomenon in which a fluid in motion within a closed duct decreases its pressure when speed increases when passing through an area with a smaller section. When the increase in speed is very large, large pressure differences are produced and then, if the end of another duct is introduced at this point of the duct, a suction of the fluid from this duct occurs, which will mix with the fluid that circulates through the first conduit. The Venturi effect is explained by the Bernoulli Principle and the continuity of the mass (Fig. 2.7).

2.3 Authors of the Navier–Stokes Equations

2.3.1 Claude-Louis Marie Henri Navier

Claude-Louis Marie Henri Navier (1785–1836) was a French engineer and physicist, a disciple of Fourier. He worked in mathematics applied to engineering, elasticity, and fluid mechanics. He is one of the 72 scientists whose name is inscribed in the Eiffel Tower (Fig. 2.8).

Among his main contributions, he is the creator of the general theory of elasticity (1821), he wrote several memories about navigation canals (1816), and he also became a railway specialist. However, his greatest contribution is the equations which describe the dynamics of an incompressible fluid, currently known as the Navier–Stokes equations.

It is also the forerunner of the calculation of structures using its hypothesis: the flat sections and perpendicular to the beam axis before deformation, remain flat and perpendicular to the beam axis after deformation. Without this contribution, sciences such as Materials Strength and Structural Calculation would not have been developed.

In 1824, he entered the French Academy of Sciences. In 1830, he was appointed professor at the National School of Bridges and Roads, and in 1831 he became professor of analysis and mechanics at the Polytechnic School of Paris, replacing Cauchy, who had to go into exile in Turin.

Fig. 2.8 Claude-Louis
Navier bust (Creative
Commons License (CC0):
public domain)

2.3.2 George Gabriel Stokes

George Gabriel Stokes (1819–1903) was an Irish mathematician and physicist, who made important contributions to fluid dynamics (Navier-Stokes equations), optics, and mathematical physics (Stokes' Theorem). He was secretary and then president of the Royal Society of United Kingdom (Fig. 2.9).

George Stokes enrolled in 1837 at Cambridge University, where 4 years later, after graduating with the highest honors (Senior Wrangler and Smith Prize), he was chosen for a teaching position. George Stokes occupied this position until 1857, when he was forced to renounce it for having married. However, 12 years later he was able to return after the statutes of the faculty were amended. He would occupy this position until 1902, when he was promoted to the mastership of his faculty.

In 1849, he became the Lucasian Professorship of mathematics at Cambridge University. Sir George Stokes, who was appointed baronet in 1889, also served his university representing it in parliament from 1887 to 1892, as one of two members of the Cambridge University Constituency. During part of this period (1885–1890), he was president of the Royal Society of which he had been secretary since 1854. As he was Lucasian Professor at the same time, he became the only person to hold these positions simultaneously because Sir Isaac Newton had done it before but not at the same time.

Stokes contributed greatly to the progress of mathematical physics. Shortly after being elected as Lucasian Professor, he announced that he considered it his professional duty to assist any member of the university with any mathematical problems.

Fig. 2.9 George Gabriel Stokes portrait (Creative Commons License (CC0): public domain)

The help provided was so real that the students had no problem consulting him about the mathematical and physical problems that were causing them difficulties.

Later, during the 30 years in which he served as secretary of the Royal Society, he also exerted an enormous influence on the advancement of the mathematical and physical sciences, not only directly by his own investigations, but also indirectly by suggesting problems in investigation and encourage people to face them. Stokes formed, along with James Clerk Maxwell and Lord Kelvin, the trio of natural philosophers, who contributed especially to the fame of the Cambridge School of Mathematical Physics in the mid-nineteenth century.

Stokes's original work began in 1840 focusing on the uniform motion of incompressible fluids. Later, he continued studying the friction of moving fluids, the balance and movement of elastic solids, and the effects of internal fluid friction on the movement of pendulums.

These investigations laid the basis of hydrodynamics and provided clues to understand natural phenomena such as cloud suspension or wave subsidence in water, as well as serving to solve practical problems such as the flow of water in rivers and canals or resistance to movement of ships.

His work in the field of fluid movement and viscosity led him to calculate what is known as Stokes' Law. The **Stokes' Law** refers to the friction force experienced by

Fig. 2.10 Balance of forces for the calculation of the sedimentation velocity

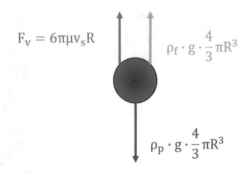

$$F_V = 6\pi\mu v_s R$$

$$\rho_f \cdot g \cdot \frac{4}{3}\pi R^3$$

$$\rho_p \cdot g \cdot \frac{4}{3}\pi R^3$$

spherical objects moving within a viscous fluid in a laminar regime of low Reynolds numbers. It was derived in 1851 by George Gabriel Stokes after solving a **particular case of the Navier–Stokes equations**. Stokes' law is only valid for the movement at low speeds of small spherical particles and has been experimentally verified for many fluids and conditions. The friction force F_v is given by Eq. 2.10.

$$F_v = 6\pi \mu v R \tag{2.10}$$

In the previous equation, R is the radius of the sphere, v its velocity, and μ the dynamic viscosity of the fluid.

If the particles fall vertically in a viscous fluid due to their own weight, their sedimentation velocity can be calculated by equating the friction force and the thrust force with the weight of the particle, as shown in Fig. 2.10.

$$v_s = \frac{2}{9} \cdot \frac{R^2 g(\rho_p - \rho_f)}{\mu} \tag{2.11}$$

where v_s is the falling velocity of the particles, g is the gravitational acceleration, ρ_p is the density of the particles, ρ_f is the density of the fluid, μ is the dynamic viscosity of the fluid, and R the equivalent radius of the particle.

In honor of his work, later, the kinematic viscosity unit in the Cegesimal System of Units was named Stokes ($1 St = 1$ cm^2/s).

In this group of works about hydrodynamics, the discovery of Stokes' Theorem must also be placed, widely used in electrical theories and capable of transforming a surface integral of a rotational vector field **F** into a line integral around from the surface boundary.

$$\iint_S (\nabla \times \mathbf{F}) \cdot d\mathbf{S} = \oint_L \mathbf{F} \cdot d\mathbf{l} \tag{2.12}$$

He also made contributions to the theory of sound, such as the influence of the gas nature in the sound intensity.

Despite his extensive work, his most famous investigations are those related to the wave theory of light. He first started with some works on light aberration and theory of certain bands of the electromagnetic spectrum. In 1849, he published a long work on the dynamic theory of diffraction, in which he showed that the polarization plane must be perpendicular to the direction of propagation.

In 1852, he published his famous work on the change in the wavelength of light, in which he described the phenomenon of fluorescence, as he showed with fluorite and uranium crystal, materials that he found to be capable of converting ultraviolet radiation (invisible) in radiation with a longer wavelength, visible.

2.4 Navier–Stokes Equations

The immediate antecedent of the Navier–Stokes Equations can be found in the Euler Equations of the mid-eighteenth century, which he obtained applying the principle of conservation of mass and Newton's Second Law to the movement of a fluid.

The forces acting on a fluid can be of two kinds: those acting at a distance, such as gravity, and those acting by contact, such as pressure. Euler considered both, but did not consider another essential contact force, viscosity, which can be defined as internal friction that opposes movement.

These equations are generally obtained by means of the so-called integral formulation, but to get their differential formulation (more useful to solve Fluid Mechanics problems) it is necessary to carry out some manipulations, mainly considering that the tangential stresses are linearly related to the gradient velocity (Newton's Viscosity Law).

It was not until 1822 when Claude Navier proposed one of the greatest contributions to science, the equations later known as Navier–Stokes, in which he did account for the effects of viscosity.

Some years later, in 1842, George Gabriel Stokes arrived at the same equations as Navier using a different way and gave them the way we know them today.

The conservation of mass equation or continuity equation is

$$\frac{\partial \rho}{\partial t} + \frac{\partial (\rho u)}{\partial x} + \frac{\partial (\rho v)}{\partial y} + \frac{\partial (\rho w)}{\partial z} = 0 \qquad (2.13)$$

In the previous expressions, u, v, and w are the three components of the velocity vector **u**.

The conservation of momentum equation is a vector equation in which the viscous stresses are represented by the stress tensor of order 2, $\bar{\tau}$.

$$\frac{\partial (\rho \mathbf{u})}{\partial t} + (\mathbf{u} \cdot \nabla)(\rho \mathbf{u}) = -\nabla P + \nabla \cdot \bar{\tau} + \rho \mathbf{g} \qquad (2.14)$$

The energy conservation equation can be expressed as follows:

$$\frac{\partial \rho \left(e + \frac{|\mathbf{u}|^2}{2}\right)}{\partial t} + \nabla \cdot \left(\rho \mathbf{u} \left(e + \frac{|\mathbf{u}|^2}{2}\right)\right) = -\mathbf{u}\nabla P + \nabla \cdot (\mathbf{u} \cdot \bar{\tau}) + \mathbf{u}\rho\mathbf{g} - \nabla \cdot \dot{\mathbf{q}} + \dot{Q}_r + \dot{Q}_q$$

(2.15)

In the previous expression, e is the internal energy of the fluid, $\dot{\mathbf{q}}$ is the heat by conduction, and $\dot{Q}_r$ and $\dot{Q}_q$ refer to the heats by radiation and reaction, respectively, applied to the control volume.

In the case of compressible fluids with negligible viscosities, the expression of the Navier–Stokes Equations coincides with the Euler Equations because the dissipative components are negligible compared to the convective ones.

On the other hand, for an incompressible (constant density) and Newtonian (constant viscosity) fluid such as water, we have the following expressions for the conservation of momentum equations:

$$\rho \left(\frac{\partial u}{\partial t} + u\frac{\partial u}{\partial x} + v\frac{\partial u}{\partial y} + w\frac{\partial u}{\partial z}\right) = \mu \left[\frac{\partial^2 u}{\partial x^2} + \frac{\partial^2 u}{\partial y^2} + \frac{\partial^2 u}{\partial z^2}\right] - \frac{\partial P}{\partial x} + \rho g_x \quad (2.16)$$

$$\rho \left(\frac{\partial v}{\partial t} + u\frac{\partial v}{\partial x} + v\frac{\partial v}{\partial y} + w\frac{\partial v}{\partial z}\right) = \mu \left[\frac{\partial^2 v}{\partial x^2} + \frac{\partial^2 v}{\partial y^2} + \frac{\partial^2 v}{\partial z^2}\right] - \frac{\partial P}{\partial y} + \rho g_y \quad (2.17)$$

$$\rho \left(\frac{\partial w}{\partial t} + u\frac{\partial w}{\partial x} + v\frac{\partial w}{\partial y} + w\frac{\partial w}{\partial z}\right) = \mu \left[\frac{\partial^2 w}{\partial x^2} + \frac{\partial^2 w}{\partial y^2} + \frac{\partial^2 w}{\partial z^2}\right] - \frac{\partial P}{\partial z} + \rho g_z$$

(2.18)

2.4.1 Exact Solutions

There are some particular solutions to the Navier–Stokes Equations.

- If the nonlinear terms are null, we have the Poiseuille flow, the Couette flow, or the Stokes law mentioned above.
- Considering the solutions with the nonzero nonlinear term, there are some interesting examples such as the Jeffery–Hamel flow, the Von Karman rotary flow, or the Taylor–Green vortex, among others.

The prediction difficulties of the Navier–Stokes Equations occur at high velocities and small viscosities, when turbulent motion develops, as Osborne Reynolds studied in 1883.

The Reynolds number is used to determine if a flow is laminar or turbulent.

$$Re = \frac{\rho v D}{\mu} \tag{2.19}$$

In Eq. 2.19, ρ is the density of the fluid, μ its dynamic viscosity, v the velocity of the fluid, and D the characteristic length, which in case of ducts, is usually its diameter.

For a straight and rigid circular section duct, the Reynolds number from which the flow is considered turbulent is 2300.

Twentieth-century studies carried out mainly by Carl Oseen, Jean Leray, and Olga Ladyzhenskaya conjecture that the Navier–Stokes Equations could develop singularities due to turbulence.

Although the initial movement was completely regular, the existence of these exact solutions mentioned above does not imply that they are stable because the regularity obtained can be destroyed at some instant, from which the solution can branch.

Within that line of research, in 2017, two mathematicians from Princeton University, Tristan Buckmaster and Vlad Vicol, demonstrated that the Navier–Stokes Equations give results that sometimes make no sense [3].

For a complete representation of the physical world, solutions are sought to be smooth, but mathematically that might not always exist with the Navier–Stokes Equations. It is important to establish a definition of what is a solution because smooth solutions require that you have a vector at each point in the fluid field. Instead, there are solutions that are called weak, in which it would be only necessary to be able to calculate a vector at some points or obtain approximate values.

In 1934, the French mathematician mentioned above, Jean Leray, came up with a class of weak solutions that fulfilled the Navier–Stokes equations. What happened was that instead of working with exact vectors, he took the mean value of the vectors in small portions of the fluid field [4].

So the strategy that Buckmaster and Vicol want to employ is to demonstrate that those weak Leray solutions are smooth. To solve the millennium problem, it would only be necessary to demonstrate that, in addition, Leray's solutions are unique. Otherwise, it would mean that for the same initial conditions the same fluid could lead to two different physical states, which makes no physical sense and would indicate that these equations do not describe reality.

Buckmaster and Vicol have only worked with weaker solutions than Leray's, but they have shown that the solutions they have used are not unique. For example, if they start with a fluid totally at rest they reach two possible situations:

- Everything was calm and remains calm forever.
- The fluid was calm, but suddenly undergoes a kind of explosion and returns to calm.

Pending future work, the only possible mathematical resolution at the moment is numerical.

2.4.2 Numerical Resolution

Computational Fluid Mechanics (CFD) is the science in charge of finding a numerical solution of the equations that govern fluid flow in a spatial and temporal domain, the Navier–Stokes equations.

It was born in the 1960s and uses computers to solve the problems of Fluid Mechanics.

The following advantages that its use entails can be distinguished:

- Lower economic cost than experimental analysis.
- Possibility of verifying theoretical results (ideal flow, 2D) impossible to validate experimentally.
- Provides complete 3D information on the field of velocities, pressures, and other dependent variables.

Among its drawbacks are:

- The reliability of the results is linked to the correct mathematical formulation of the process to be simulated.
- Long calculation time.

Its application to the industry began in the 1980s and since 1990 its expansion has been enormous, because advances in computer media allow us to face more and more complex Fluid Mechanics problems. It is present in more and more applications:

- **Aerodynamics**: air flows around buildings, aircraft, vehicles.
- **Environment**: atmospheric dispersion of pollutants.
- **Air conditioning**: heating and air renewal inside buildings, operating room ventilation.
- **Power generating equipment**: internal combustion engines, turbomachines.
- **Hydraulic installations**: flows through pumps, turbines, diffusers, valves, pipes, etc.
- **Thermal analysis**: flows in heat exchangers, vehicle radiators.
- **Medicine**: blood flow in arteries, veins and heart, respiratory flow.
- **Chemical Industry**: reactors, columns, nanotechnology.

The current market is dominated by four codes based on finite volume methods: ANSYS FLUENT (the most widely used), FLOW3D, STAR-CD, and Open FOAM.

It is necessary to carry out a validation of the CFD model with experimental measures that confirm its behavior. Validation can be in vitro, in a controlled environment, generally using the PIV technique or in vivo. In principle, the latter reflects reality better, but pressure boundary conditions cannot always be obtained and the spatiotemporal resolution of the flow velocity measurement may be insufficient.

References

1. García Sosa J, Morales Burgos A, Escalante Triay EJ (2004) Mecánica de Fluidos: antecedentes y actualidad. Ediciones de la Universidad Autónoma de Yucatán, México
2. Liñán Martínez A (2009) Las ecuaciones de euler de la mecánica de fluidos. In Alberto Galindo Tixaire and Manuel López Pellicer, editors, La obra de Euler : tricentenario del nacimiento de Leonhard Euler (1707-1783), pp 151–177. Instituto de España, Madrid
3. Buckmaster and Vicol (2019) Nonuniqueness of weak solutions to the navier-stokes equation. Ann Math 189(1):101
4. Vázquez JL (2004) La ecuación de Navier-Stokes. Un reto físico-matemático para el siglo XXI, vol 26, pp 31–56. Monografías de la Real Academia de Ciencias Exactas, Físicas, Químicas y Naturales de Zaragoza

Chapter 3
Fluid-Mechanical Description of Blood Flow

3.1 Cardiovascular System

The circulatory system constitutes the body's transportation system. It is a link, direct and indirect, between each individual cell and the homeostatic organs. The proper functioning of the cardiovascular system leads to a successful functioning of each of the other homeostatic systems (respiratory, digestive and urinary systems).

The primary function of this system is to supply oxygen and essential nutrients to the cells and to collect metabolic wastes, which are eliminated later by the kidneys, in the urine and by the air exhaled in the lungs, rich in carbon dioxide.

This system also performs other auxiliary functions. It helps endocrine system to transport the substances secreted by the glands, it carries out a body process that protects against infections and it has a role in regulating blood pressure and temperature.

The cardiovascular system is made up of the heart, which acts as an aspirating and impeller pump, and a vascular system composed of arteries, veins and capillaries, thus forming a functional unit which serves the blood, which must be in constant circulation, to irrigate the tissues.

3.1.1 Heart

The heart is an organ that has cavities, enclosed in the thoracic cavity, in the center of the chest in a place called mediastinum (mass of tissues located between the sternum and the spinal column). The blood flow in the heart, due to its characteristic velocities and sizes, is highly determined by inertial forces (Re > 2500) [1].

The original version of this chapter was revised: The author's last name in Refs. 3, 4, 9 of this chapter were incorrect. The author's last names are corrected in References. The correction to this chapter is available at https://doi.org/10.1007/978-3-030-60389-2_6.

© The Editor(s) (if applicable) and The Author(s), under exclusive license
to Springer Nature Switzerland AG 2021, corrected publication 2021
A. Pozo Álvarez, *Fluid Mechanics Applied to Medicine*,
SpringerBriefs in Computational Mechanics,
https://doi.org/10.1007/978-3-030-60389-2_3

The Pericardium is the membrane that surrounds the heart and protects valves, in relation to the chambers, and the cardiac arteries and veins that carry blood to the heart tissue. It prevents the heart from moving from its position in the mediastinum, while allowing sufficient freedom of movement for rapid and strong contraction. It consists of two main parts, the fibrous and serous pericardium.

- **The fibrous or parietal pericardium** is the superficial one and is made up of dense and resistant connective tissue. It prevents excessive stretching of the heart, protects it and fixes it to the mediastinum.
- **The serous or visceral pericardium** constitutes the deep portion. It is a thinner and more delicate membrane that forms a double layer around the heart. The outer parietal layer of the serous pericardium fissures with the fibrous pericardium. Its internal visceral layer, also called epicardium, is inserted into the surface of the heart. Between these two layers there is a thin film of serous liquid. This pericardial fluid is a slippery secretion from pericardial cells that reduces friction between membranes resulting from cardiac movements. The space that contains the pericardial fluid is the pericardial cavity.

Layers of the heart wall

The heart wall is composed of three layers: epicardium, myocardium and endo-cardium:

- **The external epicardium**, also called the visceral layer of the serous pericardium, is made up of mesothelium and thin connective tissue, which gives a smooth and slippery texture to the external surface of the heart.
- **The myocardium** or middle layer, also called the heart muscle, is the muscle layer of the heart and consists of interlocking bundles of cardiac muscle fibers. It is the most voluminous and explains the heart's ability to contract.
- **The endocardium** lines the heart chambers, covers the valves, and continues the lining membrane of the large blood vessels.

The heart is divided into two halves by a tissue wall, or partition, that runs its entire length. The right half always contains oxygen-poor blood, coming from the superior and inferior vena cava, while the left half of the heart always has oxygen-rich blood. The coming blood from the pulmonary veins will be distributed through the ramifications of the great aorta artery to oxygenate the body's tissues.

Each half of the heart has a superior cavity, the atrium, and another inferior or ventricle, with highly developed muscular walls.

There are, therefore, two atria: right and left, and two ventricles: right and left (Fig. 3.1).

- **Right atrium**
  It is a thin-walled chamber that receives blood from all parts of the body except the lungs. Three large veins empty into it: the superior vena cava, which brings venous blood from the upper body; the inferior cava, which brings venous blood from the lower body, and the coronary sinus, which drains blood from the heart itself. The right atrium pumps deoxygenated (bluish) venous blood into the right ventricle. Blood flows from the right atrium to the right ventricle through the tri-

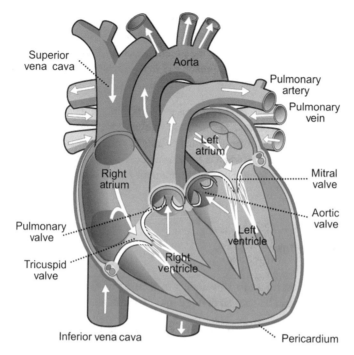

Fig. 3.1 Blood trajectory through the heart chambers (Creative Commons License (CC0): public domain)

cuspid valve, consisting mainly of fibrous tissue and so named because it consists of three leaflets or cusps.

- **Right ventricle**

  It is part of the anterior chamber of the heart. Its interior contains a series of ridges, which are formed by the protruding bundles of myocardial fibers, the fleshy tra-beculae, some of which contain most of the conduction system and nerve impulses of the heart. The interventricular septum is the division that separates the right ven-tricle from the left. Blood flows from the right ventricle through the pulmonary semilunar valve to a large artery, the trunk of the pulmonary artery, which is divided into the right and left pulmonary arteries. This chamber must be powerful in pro-pelling blood through the thousands of capillaries in the lungs and back into the left atrium of the heart.

- **Left atrium**

  It forms a large part of the heart base. It receives the already oxygenated blood, coming from the lungs, through the four pulmonary veins. After being received in this chamber, blood is pumped into the left ventricle through the mitral (or bicus-pid) valve, which has only two thicker cusps because the left ventricle is the one with the greatest pumping potential.

- **Left ventricle**

  It is the most muscular camera. Its walls are three times thicker than those of the right ventricle. With its powerful pumping, this chamber pumps blood through the aorta to all parts of the body except the lungs. Blood returns to the heart through the right atrium. Blood passes from the left ventricle through the semilunar aortic valve to the body's largest artery, the ascending aorta. From this, a portion flows to the coronary arteries, which branch off from the aorta and carry blood to the aortic arch and descending aorta.

The mitral and tricuspid valves prevent retrograde blood flow from the ventricles to the atria during systole, and the semilunar valves (aortic and pulmonary valves) prevent retrograde flow from the aortic and pulmonary arteries to the ventricles during diastole. These valves close when a retrograde pressure gradient pushes the blood back, and open when a forward pressure gradient forces the blood in an antegrade direction Hall2010.

3.1.2 Circulation

The term circulation refers to movements in a circle or along a circular path. The circulatory system can be studied divided into two smaller circulatory circuits [2].

- **Minor or pulmonary circulation**: this circuit carries blood from the heart to the lungs and from these to the heart. More specifically, blood travels from the right ventricle through the pulmonary artery to the lungs, the pulmonary arteries rapidly divide into capillaries surrounding the air sacs (alveoli), to exchange oxygen and carbon dioxide. Gradually, the capillaries gather together taking on the characteristics of veins. The veins unite to form the pulmonary veins, which carry oxygenated blood from the lungs to the left atrium.
- **Major or systemic circulation**: this circuit is the main one of the circulation. It carries oxygenated blood from the heart to all regions of the body except the lungs, and then back to the heart. All systemic arteries empty into the inferior or superior vena cava, which in turn flow into the right atrium (Fig. 3.2).

Steps through a complete circuit in the cardiovascular system:

1. **Filled left ventricle with oxygenated blood**. Blood is oxygenated in the lungs and returns to the left atrium through the pulmonary vein. This blood then flows from the atrium to the left ventricle through the mitral valve.
2. **From the left ventricle blood is expelled into the aorta**. Blood leaves the left ventricle through the aortic valve located between the left ventricle and the aorta. When the left ventricle contracts, the pressure inside the ventricle increases, causing the opening of the aortic valve and blood to flow into the aorta. The blood then flows through the arterial system driven by the pressure generated by the contraction of the left ventricle.
3. **Cardiac output is distributed among different organs**. The total cardiac output of the left hemicardium is distributed to the organs of the body by means of groups of parallel arteries. Thus, blood flow is simultaneously provided to the

Fig. 3.2 Blood distribution in the circulatory system

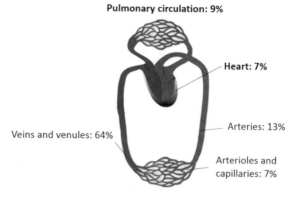

Pulmonary circulation: 9%

Heart: 7%

Arteries: 13%

Veins and venules: 64%

Arterioles and capillaries: 7%

Systemic circulation: 84%

brain through the cerebral arteries, to the heart through the coronary arteries, and to the kidneys through the renal arteries.

4. **Blood flow from the organs is collected in the veins**. The blood leaving the organs contains waste products of metabolism, such as carbon dioxide. This mixed venous blood is collected in veins of increasing diameter and finally in the major vein, the vena cava. This carries the blood to the right hemicardium.

5. **Venous return to the right atrium**. The pressure in the vena cava is greater than that of the atrium. Consequently, the latter fills with blood (venous return).

6. **Mixed venous blood fills the right ventricle**. Mixed venous blood flows from the right atrium to the right ventricle through the tricuspid valve in the right hemicardium.

7. **Blood is expelled from the right ventricle to the pulmonary artery**. When the right ventricle contracts, blood is expelled through the pulmonary valves into the pulmonary artery, which carries blood to the lungs. In the capillary bed of the lungs, oxygen is added to the blood from the alveolar gas and the carbon dioxide is removed, which is added to the alveolar gas. Thus, the blood leaving the lungs contains more oxygen and less carbon dioxide.

8. **Blood from the lungs returns to the heart through the pulmonary veins**. Oxygenated blood returns to the left atrium through the pulmonary vein for a new cycle.

3.1.3 Arterial, Venous and Capillary Systems

3.1.3.1 Arterial System

Arteries are vessels through which blood circulates from the heart's ventricles to the tissues with the required oxygen and nutrients. Large-caliber elastic arteries arise in the heart (aorta), the diameter of which can be as much as 3 cm. They branch into smaller diameter, thinner walled muscle arteries as they move away from the heart. These muscular arteries are divided in turn into smaller ones, the arterioles, and can reach 0.2 mm in diameter. When they enter the tissues, they branch into countless microscopic vessels, known as capillaries. The wall of the arteries has three layers or tunics: internal, intermediate and external tunic.

The walls of some of the arteries and arterioles have, in addition to their elastic tunic, a muscular tunic. Due to elastic fibers, arteries tend to have high compliance, which means that their walls stretch or expand without tearing in response to small increases in pressure. Therefore, they are able to maintain their cylindrical shape even when they are empty, contrary to what happens in the veins. Two arterial trunks leave the base of the heart: the pulmonary artery, from the right ventricle and the aorta artery, from the left ventricle.

Larger diameter arteries are called elastic arteries because their middle layer contains a high proportion of elastic fibers and their walls are relatively thin relative to their diameter. These arteries carry blood from the heart to those of an intermediate caliber, which are more muscular because their middle tunic contains more smooth muscles and fewer elastic fibers than the conduction arteries. In them, vasoconstriction and vasodilation are possible to a greater degree, to regulate blood flow, these arteries are also known as distribution arteries.

The arteriole is a very small diameter duct, the result of the division of an artery, which distributes blood to the capillaries of an artery. When arteriolar smooth muscle contracts with consequent vasoconstriction, the blood flow in the capillaries decreases, and when it relaxes, vasodilation increases this flow. In the arterioles there is a great resistance to the passage of the flow, which causes a high pressure drop, which causes them to be subjected to high levels of pressure. In the arterial system, the viscous forces are the dominant ones, (Re < 500), due to the characteristic velocities and diameter of these blood vessels [1].

3.1.3.2 Venous System

Veins and venules are thin-walled, inelastic blood vessels prepared to return blood to the atria of the heart. Despite the fact that the veins are made up of essentially the same 3 layers (tunics) as the arteries, the relative thickness of the layers is different. The internal tunic of the veins is thinner than that of the arteries. The middle tunic of the veins is much thinner and contains fewer elastic fibers than the arteries. The outer tunic of the veins is the thickest layer and is made of elastic fibers and collagen.

The diameters in the venous system are of the same order of magnitude (slightly smaller) than in the arterial system, so inertial forces can again be influential. However, both velocities and pressure are less than in the arterial system. As a consequence, non-stationary inertial forces are less important than in the arterial system. Also the pressure between the veins is significantly lower, close to 1 kPa.

Due to the low pressures, the gravitational forces become important, especially in the vertical position. To overcome the effect of gravity, veins have valves that prevent reverse flow and ensure one-way flow to the heart.

3.1.3.3 Capillary System

Capillaries are arteries that go into the tissues and branch into countless microscopic vessels that connect arterioles to venules. These reach almost every cell in the body and their main function is to allow the exchange of nutrients and waste between the blood and the cells of the tissues through the interstitial fluid.

The blood flow from arterioles to venules through the capillaries is called microcirculation. The diameter of the capillaries is small enough (between 4 and 10 μm) that the blood can no longer be considered as a homogeneous fluid. The capillary walls are composed only of a layer of endothelial cells and a basement membrane.

In the capillary system, viscous forces dominate over inertial forces, Re<1. Consequently, the microcirculation, including the arterioles, can be approximated as a collection of parallel tubes or a porous medium that generates a loss of pressure in the flow between the arteries and the veins that is known as perfusion pressure [1].

$$P_a - P_v = R_p \cdot Q \tag{3.1}$$

In this linear relationship, P_a is the arterial pressure, P_v the venous pressure, Q the blood flow rate and R_p is the peripheral resistance, mainly controlled by the muscle fibers of the arterioles.

Velocities and Reynolds numbers shown in the Table 3.1 correspond to average values.

Table 3.1 Summary of the characteristics of the vessels of the circulatory system [3]

Vessel	Number	Diameter (cm)	Thickness (cm)	Velocity (m/s)	Reynolds	Womersley
Aorta	1	2.5	0.2	0.48	3400	18
Arteries	159	0.4	0.1	0.45	500	3
Arterioles	5.7×10^7	5×10^{-3}	2×10^{-3}	0.05	0.7	0.035
Capillaries	1.6×10^{10}	8×10^{-4}	1×10^{-4}	1×10^{-3}	0.002	0.006
Venules	1.3×10^9	2×10^{-3}	2×10^{-4}	2×10^{-3}	0.01	0.015
Veins	200	0.5	0.05	0.1	140	4
Vena cava	1	3	0.15	0.38	3300	21

3.1.4 The Blood

Blood is a fluid of complex composition. Its main function is to provide nutrients to the body's cells (oxygen, glucose) and transport waste (carbon dioxide, lactic acid). It also allows transporting different substances (amino acids, hormones, lipids) between organs and tissues. Blood represents approximately 7% of human weight, equivalent to about 5 L in an adult.

Blood is a liquid, made up of water and dissolved organic and inorganic substances (mineral salts), which form the blood plasma constituting 60% of the total volume and three types of elements or blood cells: red blood cells, white blood cells and platelets, occupying 40% of the total volume [4] (Table 3.2).

Blood plasma is the liquid part of the blood, mostly water. It is salty, yellowish in color and floats the other components of the blood, it also carries food and waste substances collected from the cells.

Red blood cells are responsible of the distribution of molecular oxygen. Red blood cells have a reddish pigment called hemoglobin that helps them carry oxygen from the lungs to the cells. Insufficient amount of hemoglobin or red blood cells in the body leads to anemia.

White blood cells or leukocytes play an important role in the Immune System when cleaning (phagocytes) and defense (lymphocytes) work. Platelets are very small fragments of cells that serve to plug wounds and prevent bleeding.

Table 3.2 Approximate composition of the blood indicating constituent elements [5]

Cellular part (40%)		
Cell type	Cell concentration	Characteristic size
Red blood cells (99.7%)	5000000/µL	Biconcave discs (8 µm of diameter and 2.5 µm of thickness
White blood cells (0.2%)	7500/µL	Spherical shape (diameter from 20 to 100 µm)
Platelet (0.1%)	250000/µL	Ellipsoidal shape (4 µm major axis and 1.5 µm minor axis)
Plasma part (60%)		
Composition	Main element	Function
Water (92%)	H_2O	Reduce viscosity
Plasma proteins (7%)	Albumin (60%)	Osmotic pressure
	Globulin (35%)	Immune function
	Fibrinogen (3%)	Coagulation
	Others (2%)	Enzymes/Hormones
Other solutes (1%)	Electrolytes	Homeostasis
	Nutrients	Cellular energy
	Waste	Excretion

The amount of solid particles suspended in the blood offers resistance to the advance of the flow. This phenomenon causes that the blood has a higher viscous behavior than water. Furthermore, given the large amount of water present in the blood, it has incompressible properties. The fluid that simulates its behavior must be incompressible and have an adequate viscosity.

3.1.4.1 Density

Blood is an incompressible fluid due to the large amount of water it has. This means that the density of the blood remains constant with pressure changes, although it may vary from person to person depending on their metabolism. It is usually considered a constant value of 1055 kg/m^3 [4].

3.1.4.2 Dynamic Viscosity

Blood is a non-Newtonian fluid. Despite the fact that plasma behaves like a Newtonian fluid, the presence of red blood cells (hematocrit), which account for up to 99% of the particles present in the blood, do the viscosity variable. This is modified depending on the dimensions of the tube and the type of flow. When the blood velocity increases, the viscosity decreases. The viscosity of the blood depends drastically on the hematocrit, increasing its value when the hematocrit percentage is higher. For normal physiological blood levels of around 40% hematocrit percentage, a mean blood viscosity value of 3.5 centipoise (0.0035 Pa·s) is considered at the temperature of 38 °C [4].

3.2 Cardiac Cycle

Cardiac events that occur from the beginning of one heart beat to the beginning of the next are called the cardiac cycle. Due to the arrangement of the conduction system there is a delay of more than 0.1 s in the passage of the cardiac impulse from the atria to the ventricles. Blood normally flows continuously from the large veins into the atria, where approximately 80% of the blood flows directly into the ventricles, even before the atria contract. Thereafter, atrial contraction usually results in an additional 20% filling of the ventricles. Thus, the atria act as priming pumps for the ventricles, and the ventricles in turn provide the primary source of power to propel blood through the vascular system [2].

The cardiac cycle is composed of a relaxation period called diastole, followed by a contraction period called systole, as it can be seen in Fig. 3.3. The cardiac cycle period is the inverse value of the heart rate. For a heart rate of 75 beats per minute the duration of the cardiac cycle is 0.8 s. The top three curves in Fig. 3.3 correspond to pressure changes in the aorta, left ventricle, and left atrium, respectively. Although

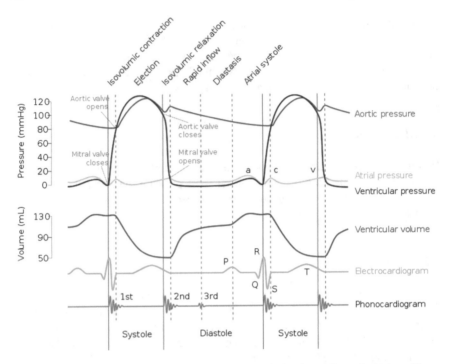

Fig. 3.3 Pressure-Volume changes in the left hemicardium during the cardiac cycle (Adapted from: https://en.wikipedia.org/wiki/File:Wiggers_Diagram_2.svg)

the atrial pressure remains more or less constant, three small increases in pressure are observed (a, c and v) [2]. The fourth curve represents changes in left ventricular volume, and the fifth represents the electrocardiogram (ECG).

3.2.1 Cardiac Cycle Phases

Seven phases of the cardiac cycle can be distinguished, explained below for the left heart hemicardium. The same events occur in the right hemicardium, with the difference that the pressure in the left hemicardium is significantly higher [6].

- **Phase 1: Atrial contraction**. This is the first phase of the cardiac cycle and begins with the **P wave** corresponding to the electrocardiogram. This wave is produced by the propagation of depolarization in the atria and has a duration of 50 ms approximately [7]. It corresponds to the last part of ventricular diastole, in which all the chambers are relaxed and the left ventricle is partially filled with blood. There is hardly any flow through the mitral valve at this stage because left atrium and left ventricle pressures are practically the same. When the atrium contracts, the pressure inside it increases, causing additional blood flow to the left ventricle, corresponding to **a wave** in the diagram. Usually the right atrial pressure increases from 4 to 6 mmHg during atrial contraction and the left atrial pressure increases from approximately 7 to 8 mmHg [2]. The pressure in the left atrium exceeds the pressure in the veins, but only small amounts of reverse flow take place. Atrial systole lasts for about 100 ms.

- **Phase 2: Isovolumic contraction**. This is the first stage of ventricular systole. Approximately 150 ms after the start of the P wave, **QRS waves** (90 ms of duration) appear as a consequence of electrical depolarization of the ventricles, which initiates contraction of the ventricles and an elevation of ventricular pressure [7]. As the pressure inside the ventricle exceeds the atrial pressure, the mitral valve closes immediately. The ventricle is now a closed chamber. For a short period of time the pressure continues to build rapidly while valves are closed. During this phase there are no changes in volume, that is why it is said to be isovolumetric. The rapid increase in left ventricular pressure causes the mitral valve to buckle into the left atrium, this can be seen in a small spike in the atrial pressure curve, the **c wave**.
- **Phase 3: Rapid ejection**. Once the pressure in the left ventricle exceeds the pressure in the aorta (approximately 80 mmHg), the aortic valve opens and a rapid ejection of blood into the aorta begins. The ventricular muscles begin to shorten, and the volume of the ventricle decreases. The pressure gradient between the aorta and the left ventricle is very small due to the large opening of the aortic valve (low resistance). As a result of contraction and shortening of the left ventricle, the mitral annulus descends and the left atrium expands slightly with a pressure drop occurring in the left atrium. Venous blood continues to pass into the left atrium, and the atrial pressure increases again.
- **Phase 4: Reduced ejection**. **T wave** is observed in the electrocardiogram, representing the repolarization of the ventricles, when the fibers of the ventricular muscle begin to relax. This wave lasts approximately 50 ms and occurs 250 ms after the QRS complex ends [7]. The pressure in the left ventricle is reduced and the period of reduced ejection begins. The pressure in the left ventricle gradually decreases and falls slightly below the pressure in the aorta, which is also decreasing. Instead, blood continues to leak from the left ventricle due to inertia. At the end of systole, the pressure in the left ventricle drops faster and reverse blood flow to the left ventricle appears. Blood flows to the cusps of the aortic valve, which closes abruptly. Passive filling of the atrium continues during this period until the end of the fifth phase.
  For a normal person at rest, ventricular systole (phases 2, 3 and 4) usually lasts about 270 ms [8].
- **Phase 5: Isovolumic relaxation**. After the aortic valve closes, the ventricle continues to relax and the pressure drops dramatically. The left ventricle volume remains constant because all the valves are closed. This is the beginning of ventricular diastole. The pressure in the atrium now reaches its maximum, as can be seen at the peak of the **v wave**.
- **Phase 6: Rapid filling**. When the pressure in the left ventricle falls below the pressure in the left atrium, the mitral valve opens rapidly. The blood accumulated

in the atrium during systole now flows into the left ventricle. Pressure in the left ventricle and left atrium continues to drop, that of the atrium because it is emptying into the ventricle and that of the ventricle because it is still under relaxation. Ongoing relaxation of the left ventricle creates additional suction of blood from the left atrium. The volume in the left atrium decreases while the left ventricle is expanding.

- **Phase 7: Reduced filling**. As the left ventricle continues to fill and expand, the pressure in the left ventricle begins to increase again. This reduces the pressure gradient between the two chambers and the filling slows down. During this period the pulmonary veins fill the left atrium and restore a positive atrioventricular pressure gradient.

Ventricular diastole comprises phases 5, 6, 7 and 1 and lasts approximately 530 ms for a person with a heart rate of 75 beats per minute.

3.2.2 Heart Work Generation

The systolic heart work is the amount of energy that converts to work during each beat while it pumps blood into the arteries. The heart work is used in two ways. First, in greater proportion, to move blood from the low-pressure veins to the high-pressure arteries (volume-pressure work or external work). Second, a small proportion of energy is used to accelerate the blood to its ejection rate through the aortic and pulmonary valves (kinetic energy of the blood flow).

The external work of the right ventricle is normally one sixth of the work of the left ventricle due to the six-fold difference in systolic pressures in both ventricles. Typically, the left ventricular work required to create kinetic energy for blood flow is only 1% of total ventricular work and it is ignored calculating total systolic work. But in abnormal situations, such as aortic stenosis, in which blood flows at high velocities through the stenosed valve since the passage section has been reduced, it may take more than 50% of the total work to generate the kinetic energy of blood flow [2] (Fig. 3.4).

The area enclosed by the pressure-volume diagram of the ventricle represents the net external cardiac work during its contraction cycle. When the heart pumps large amounts of blood during periods of increased activity, the area of this diagram increase. The pressure-volume diagram of the left ventricle is divided into four phases:

- **Phase I: Filling period**. This phase begins with a ventricular volume of approximately 50 mL from the previous beat (**end systolic volume**) and a diastolic pressure close to 2–3 mmHg. As venous blood flows into the ventricle from the left atrium, the ventricular volume increases to about 120 mL, the so-called **end-diastolic volume**. Diastolic pressure increases to approximately 5–7 mmHg.

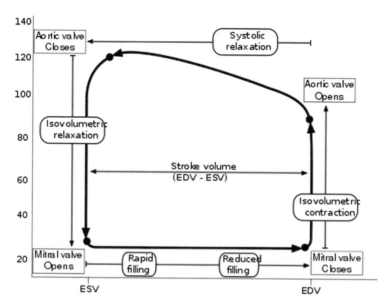

Fig. 3.4 Pressure volume loop of the cardiac cycle

- **Phase II: Period of isovolumic contraction**. The volume of the ventricle is unchanged because all the valves are closed. However, the pressure inside the ventricle increases until it equals the pressure in the aorta, approximately 80 mmHg.
- **Phase III: Ejection period**. During ejection, systolic pressure increases even more due to more intense contraction of the ventricle. At the same time, the volume of the ventricle decreases because the aortic valve has already been opened and blood leaves the ventricle into the aorta. The **total ejection volume** is calculated as the difference between the end-diastolic and the end-systolic volume, and is usually around 70 mL [2].
- **Phase IV: Isovolumometric relaxation period**. At the end of the ejection period, the aortic valve closes, and the ventricular pressure drops again to the level of diastolic pressure with no volume changes. Thus, the ventricle recovers its initial value, with approximately 50 mL of blood in the ventricle and a diastolic pressure of 2 to 3 mmHg.

3.3 Fluid-Dynamic Characterization of Blood Flow

The branch of biomechanics that is in charge of studying the blood flow inside all the blood structures (heart, arteries, veins, capillaries) based on the physical principles of fluid mechanics is hemodynamics.

3.3.1   Hemodynamics History

The following is a historical overview of how attempts have been made to describe the fluid dynamic behavior of the circulatory system to this day.

- In the 4th century BC Aristotle described the communication of the heart with the blood vessels.
- In the 3rd century BC Praxagoras distinguished the functions of the veins and arteries, noting that the arteries powered blood and the veins did not.
- At the end of the 2nd century, the Greek Galen proposed that pressure pulses originating in the heart propagate through the arterial network.
- At the beginning of the 17th century, the Englishman William Harvey, through studies where he measured the flow rate of blood passing through the veins, concluded that the blood flow is unidirectional and that the blood recirculates continuously.
- In the middle of the 17th century, Malpighi and van Leeuwenhoek discovered the capillaries, which linked veins with arteries. This confirmed that the circulatory system is closed and the flow is unidirectional.
- In the early 18th century, Stephen Hales made the first blood pressure measurements in live animals, observing that it was pulsatile. He started the study of hemodynamics, he wanted to investigate how pressure was transmitted through blood vessels.
- In the 18th century, it was found that the circulatory system has the properties of storing energy due to the elasticity of the arterial walls and that it dissipates energy thanks to the viscous behavior of the blood.
- In the late 18th and early 19th centuries, Young described the relationship between the elastic properties of arteries and the propagation velocity of the pulse wave.
- In the 19th century, the French Poiseuille established the law that bears his name:

$$Q = \Gamma \cdot \frac{\Delta P \cdot D^4}{\mu \cdot L} \tag{3.2}$$

Q is the flow, D the diameter of the duct, L the length, ΔP the pressure difference between the ends of the duct, μ the dynamic viscosity and Γ is a dimensionless parameter that depends on the geometry (for circular section $\Gamma = \pi/32$). He also deduced the parabolic velocity profile in the section.

- In the same century, Moens determined empirically the transmission velocity, c of a pressure wave in a thin-walled elastic tube with incompressible, non-viscous fluid:

$$c = \sqrt{\frac{E \cdot h}{\rho_f \cdot D}} \tag{3.3}$$

- Despite the achievements got in previous centuries, it was not until the 20th century when experimental and mathematical methods emerged, which made it possible

to describe the pulsatile nature of blood flow. At the beginning of the 20th century, the Windkessel models were established, where the aorta was conceived as an elastic tube with the capacity to store fluid. At one end of the heart the fluid is intermittently introduced, while at the other end the fluid comes out approximately constantly. The circulatory system is conceived as an elastic reservoir where the heart pumps blood and from which a network of non-elastic ducts exits to irrigate the body. The resistance of the flow would be given by Poiseuille's Law.
- In the 1950s, Womersley provided the first fundamental analytical solution for the equations that govern blood flow. For this, it considered only the linear terms of the Navier-Stokes equation. He derived an equation that predicts flow under a sinusoidal function of pressure in a rigid tube. Subsequently, it was resolved for elastic walls and with new contour conditions closer to reality.

3.3.2 Hagen–Poiseuille Law

It establishes the analytical solution of a steady laminar flow of an incompressible fluid through a straight and rigid duct of constant section, subjected to a pressure difference at its ends.

The sum of the pressure p, and the potential energy U, at each point is called reduced pressure.

For a duct of length , the reduced pressure gradient Pe is equal to the difference of the reduced pressures at the inlet i, and outlet f, of the duct divided by its length.

$$Pe = -\frac{\partial(p + \rho \cdot U)}{\partial x} = \frac{(p + \rho \cdot U)_i - (p + \rho \cdot U)_f}{L} \tag{3.4}$$

The differential equation of motion (Eq. 3.4) is solved in cylindrical coordinates for a duct of circular section of radius R with the following boundary condition: $u(r = R) = 0$

$$0 = Pe + \mu \cdot \frac{1}{r} \cdot \frac{\partial}{\partial r}\left(r \cdot \frac{\partial u}{\partial r}\right) \tag{3.5}$$

The resulting velocity profile follows Eq. 3.6 and is a parabola where the velocity reaches the maximum value on the axis of the duct, reducing to 0 on the walls, as it can be seen in Fig. 3.5.

$$u = \frac{Pe}{4\mu} \cdot (R^2 - r^2) \tag{3.6}$$

The flow rate Q is obtained integrating the velocity profile for the entire section using the Eq. 3.7.

Fig. 3.5 Hagen Poiseuille
velocity profile

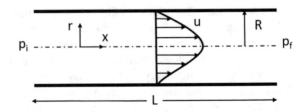

$$Q = \int_0^R 2\pi r \cdot u \cdot dr = \frac{\pi \cdot R^4}{8 \cdot \mu} \cdot Pe \qquad (3.7)$$

The Hagen-Poiseuille flow introduces numerous simplifications that deviate from the actual blood flow. However, in certain cases it can be considered as a good first approximation of the flow [5]:

- **Newtonian fluid**. Blood is actually a non-Newtonian fluid, but when it is subjected to high velocity gradients ($\frac{\partial u}{\partial r}$) its behavior is more similar to that of a Newtonian fluid, as in the arteries of more than 0.5 mm of diameter.
- **Laminar flow**. To determine if a flow is laminar or turbulent, Reynolds number, previously defined, is used. The typical range of Reynolds numbers in the blood vessels can range from 1 in the small arterioles to approximately 4000 in the aortic artery during the systolic peak [9]. This would be the most unfavorable case, although on average it does not exceed the critical Reynolds for a straight and rigid circular section duct (2300).
  However, it must be borne in mind that the blood vessels are neither straight nor rigid and the section is not perfectly circular. For that reason, the critical Reynolds for a blood vessel would be approximately 500 [10].
- **No slippage on the walls**. The layer of blood closest to the arterial wall is firmly attached to it, so this non-slip condition is entirely reasonable.
- **Steady flow**. The blood flow has a pulsating character: accelerations and decelerations appear in the flow, which are not taken into account under the assumption of steady flow. Therefore, this approach could only be interesting when it is intended to have an order of magnitude of the average flow with the consequent simplification in the mathematical approach that this entails.
- **Cylindrical geometry**. Arteries can be considered to have roughly circular cross sections, unlike veins, whose shape is more closely to an ellipse.
- **Rigid walls**. Arterial walls are flexible and deform with pressure changes throughout the cardiac cycle, whereby the interaction between flow and wall deformation has a not inconsiderable effect.
- **Flow developed**. In the bifurcations the velocity profile is not parabolic until it exceeds the detachment zone in the entrance region and it can be considered developed flow. The length of the detachment zone depends on the Reynolds number.

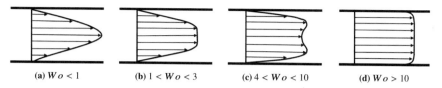

(a) $Wo < 1$ (b) $1 < Wo < 3$ (c) $4 < Wo < 10$ (d) $Wo > 10$

Fig. 3.6 Velocity profiles for different Womersley number values

3.3.3 Pulsating Flow

Blood flow is considered in most cases as laminar flow. However, it is pulsatile and under certain conditions and in small time intervals the flow can be dominated by inertial forces [5]. So another dimensionless parameter, the **Womersley number**, is used to characterize an unsteady flow. The velocity profile of a pulsating flow depends on the size of the duct D, the angular frequency of the pulse ω and the density ρ and viscosity μ of the fluid. This is reflected in the Womersley dimensionless parameter, Wo, which expresses the importance of unsteady inertial forces versus viscous forces.

$$Wo = \frac{D}{2} \cdot \sqrt{\frac{\omega \cdot \rho}{\mu}} \tag{3.8}$$

For low Womersley parameter values ($Wo < 1$), as in narrow vessels and with low frequency values, the viscous stresses will dominate and we will have a quasi-stationary parabolic velocity profile governed by Poiseuille's Law, such as in Fig. 3.6(a). For higher values ($1 < Wo < 3$) the instantaneous velocity profile lags behind the instantaneous pressure gradient. If the values are already very high ($Wo > 4$), inertia dominates the viscous effects and the velocity profile becomes practically flat, as in Fig. 3.6(d) and the flow has a marked unsteady character [11].

The Womersley flow is obtained from the Navier-Stokes equations, imposing a series of hypotheses, similar to those of Poiseuille's law for steady flow: straight, rigid and constant section duct. Assuming a sinusoidal pressure gradient of amplitude $\frac{\hat{\partial p}}{\partial z}$ and angular frequency ω, the velocity profile follows Eq. 3.9 and the flow rate $Q(t)$, Eq. 3.10 [1].

$$u(r, t) = Re \left[\frac{i}{\rho\omega} \cdot \frac{\hat{\partial p}}{\partial z} \left\{ 1 - \frac{J_0\left(i^{\frac{3}{2}} Wo \cdot y\right)}{J_0\left(i^{\frac{3}{2}} y\right)} \right\} e^{i\omega t} \right] \tag{3.9}$$

$$Q(t) = Re \left[\frac{i\pi R^2}{\rho\omega} \cdot \frac{\hat{\partial p}}{\partial z} \left\{ 1 - \frac{2J_1\left(i^{\frac{3}{2}} Wo\right)}{i^{\frac{3}{2}} Wo J_0\left(i^{\frac{3}{2}}\right)} \right\} e^{i\omega t} \right] \tag{3.10}$$

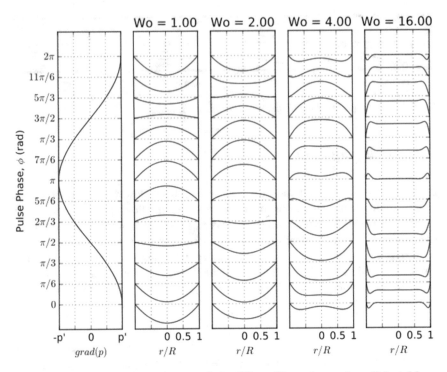

Fig. 3.7 Theoretical velocity profile in a cycle for different Womersley numbers (Adapted from: https://wikivisually.com/wiki/Pulsatile_flow)

In the above expressions i is the imaginary number ($i = \sqrt{-1}$), $y = \frac{r}{R}$ is the relative radial position, Wo is the Womersley number and J_0 and J_1 are Bessel functions of order 0 and 1 respectively. In the frequency domain the solution is analytical but in the time domain the solution can only be obtained through numerical integration.

Figure 3.7 shows the evolution of the velocity profile for different instants of time in a cycle with a sinusoidal pressure gradient. It is observed that at certain moments of the cycle, if the Womersley number is high, the velocity at the points near the wall differs from the velocities in the center of the section, where the velocity profile is practically flat.

References

1. van de Vosse FN, Stergiopulos N (2011) Pulse wave propagation in the arterial tree. Ann Rev Fluid Mech 43(1):467–499
2. Hall John E (2010) Guyton and hall textbook of medical physiology, 12e. Saunders
3. Calvo Plaza FJ (2006) Simulación del flujo sanguíneo y su interacción con la pared arterial mediante modelos de elementos finitos. Septiembre 2006

4. Vlachopulos C, O'Rourke M, Nichols WW (2011) McDonald's blood flow in arteries: theoretical, experimental and clinical principles, 6th edn. HODDER & STROUGHTON
5. Chandran Krishnan B, Rittgers Stanley E, Yoganathan Ajit P (2012) Biofluid mechanics: the human circulation, 2nd edn. CRC Press
6. Dahl SK, Thomassen E, Hellevik LR, Skallerud B (2012) Impact of pulmonary venous locations on the intra-atrial flow and the mitral valve plane velocity profile. Cardiovascular Eng Technol 3(3):269–281, Sep 2012
7. Wei Y, Frame Mary D (2011) Biofluid mechanics: an introduction to fluid mechanics, Macrocirculation, and microcirculation (Biomedical Engineering). Academic Press
8. Klabunde RE (2011) Cardiovascular physiology concepts. Lippincott Williams&Wilki
9. Ku DN (1997) Blood flow in arteries. Ann Rev Fluid Mech 29(1):399–434
10. Ferrari M, Werner GS, Bahrmann P, Richartz BM, Figulla HR (2006) Turbulent flow as a cause for underestimating coronary flow reserve measured by doppler guide wire. Cardiovascular Ultrasound 4(1)
11. Yazdi SG, Geoghegan PH, Docherty PD, Jermy M, Khanafer A (2018) A review of arterial phantom fabrication methods for flow measurement using piv techniques. Ann Biomed Eng 46(11):1697–1721

Chapter 4
Cardiac Flow Visualization Techniques

The following classification regarding cardiac flow measurement techniques can be established:

- **In vivo techniques**. Those that are carried out directly on the patient's organism, that is why they reflect best the reality a priori. They do not require direct visual access to the study area.
- **In vitro techniques**. They are performed outside the living organism. As it is not always possible to perform in vivo tests, with the in vitro techniques a much more controlled environment is achieved, which makes them ideal for the validation of numerical codes. They require direct visual access to the study region.
- **In silico techniques**. Refers to computational studies performed using numerical CFD simulation. These numerical models need an in vivo or in vitro validation.

4.1 In Vivo Techniques

4.1.1 Characterization of the Flow in the Heart

There are several in vivo techniques based on clinical images that make it possible to characterize the functioning of the heart:

- **Echocardiography**, based on the use of ultrasound. There are different types:

  - **TTE (Transthoracic echocardiography)**. It is a non-invasive technique and constitutes the most common type of echocardiogram. The ultrasonic transducer

---

The original version of this chapter was revised: The author's last name in Refs. 22, 23, 30, 33 of this chapter were incorrect. The author's last names are corrected in References. The correction to this chapter is available at https://doi.org/10.1007/978-3-030-60389-2_6.

© The Editor(s) (if applicable) and The Author(s), under exclusive license
to Springer Nature Switzerland AG 2021, corrected publication 2021
A. Pozo Álvarez, *Fluid Mechanics Applied to Medicine*,
SpringerBriefs in Computational Mechanics,
https://doi.org/10.1007/978-3-030-60389-2_4

is placed on the chest wall and varying its position and orientation, images of different planes of the heart can be obtained. This technique allows the reconstruction of 2D and 3D images.

- **TEE (Transesophageal Echocardiography)**. It is an invasive technique consisting of the introduction of an ultrasonic transducer through the patient's esophagus. This technique is used more frequently when transthoracic images are not optimal and greater precision is needed for evaluation, such as in case of heart valves. It allows obtaining 2D and 3D images.
- **ICE (Intracardiac Echocardiography)**. Provides high-resolution images from inside the heart with the introduction of a venous catheter.

- **Doppler echocardiography**. It is a variety of traditional echocardiography in which, taking advantage of the Doppler effect, it is possible to determine if the flow is directed towards or away from the transducer. The velocity component of such flow can only be determined along the transducer direction by Eq. 4.1.

$$v = \frac{(f_e - f_r) \cdot c}{2 f_e \cdot cos\theta} \tag{4.1}$$

In Eq. 4.1 v is the velocity to be estimated, f_e the emission frequency, f_r the reception frequency, θ the angle between the ultrasound beam and the object under study and c is the velocity of sound transmission in the tissues.

The angle θ must be as close to 0 (aligned ultrasound beam and study object) because the cosine would be 1 and the velocity measurement would be exact. Any greater angle would suppose an error in the measurement being impossible to measure at 90°. In practice, the largest allowed angle is 20°, which means a measurement error of 6%.

There are three types of Doppler applied to the study of blood flow [1]:

- **Pulsed Doppler**. The transducer emits an ultrasound pulse and after a certain time, depending on the depth of the measurement point, receives it. In this way, the flow velocity at the point on the line where the sample volume is located is known. The spectrum for that point is represented by a shiny black border inside. Its main limitation is the inability to measure high velocities (>1 m/s).

  The pulse repetition frequency depends on the time it takes to receive the reflected echo, so the greater the depth of the sample volume, the lower the emission frequency and therefore the maximum velocities it could detect are also lower. On the other hand, the Pulsed Doppler signal becomes saturated when the maximum velocity that is able to measure accurately for each pulse repetition frequency (Nyquist limit) is exceeded. The *Aliasing* phenomenon occurs, whereby the system mistakenly detects velocities. To avoid this, the sampling frequency must be at least twice the emission frequency. The *Aliasing* can be decreased reducing the depth to which the sample volume is located, which increases the frequency of repetition of pulses and/or changing the position of the baseline. This technique is useful for measuring velocity at a given spatial location where high velocities are not reached, such as flow through the mitral valve.

– **Continuous Doppler**. The transducer emits and receives simultaneously, so that it does not have a repetition pulse rate limitation and therefore can register high velocities (even higher than 6 m/s) as those that occur in stenotic holes. It registers the velocities in the entire ultrasound beam and not in a specific point, unlike Pulsed Doppler. The spectrum representation is bright both on the edge and inside, which means that all the velocity ranges present along the ultrasound beam are detected, not allowing to differentiate near and far fluxes, therefore, the exact point at which those high velocities are reached.

– **Color Doppler**. This is a variant of Pulsed Doppler that represents a color-coded average velocity map, superimposed on the 2D image of the cardiac anatomy. This information is obtained in the same way as with Pulsed Doppler, simultaneously interrogating multiple sample volumes sequentially on a selected surface called color box. This box is made up of a series of pixels, coded with a different color depending on the direction of the detected movement (usually red for the flows that approach the transducer and blue for those that move away) and whose color saturation is proportional to the average frequency of echoes from the volume that equals that pixel. To obtain an adequate estimate of the average frequency, the transducer must send a minimum of eight to ten pulses for each color line, affecting the number of images per second that can be displayed (*frame rate*). The color box should not cover the entire image not to decrease temporal resolution excessively.

Color Doppler enables the rapid display of large amounts of information of hemodynamic interest. However, it does not represent the maximum velocity but the average velocity. Also, *Aliasing* can appear easily as the Nyquist limit can be reached for flows with speeds not excessively high (0.6 6 m/s) [2] (Fig. 4.1).

- **Echo-PIV (Echocardiographic Particle Image Velocimetry)**. Technique based on cross-correlation of particle image fields acquired by ultrasound. It is derived from optical PIV, which is a usually 2D in vitro technique but can allow 3D visualization of the flow by combining measurements in multiple planes, so it does not have the limitations of Doppler. Multiphase bubbles of a gas with a hydrophilic envelope are used as contrast agents. The size of these bubbles must coincide with that of the red blood cells to avoid blockage of the pulmonary capillaries when

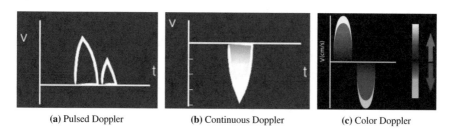

(a) Pulsed Doppler   (b) Continuous Doppler   (c) Color Doppler

Fig. 4.1 Types of Doppler Echocardiography

administered intravenously. Inside the arteries and the heart, these bubbles create aggregates that generate great dispersion when under an ultrasound field. Echo-PIV uses these aggregates as tracer particles [3].

A limitation of this technique is related to the complex interaction of ultrasound with bubble aggregates. This causes the tracers to disappear from the image plane, not only because of out-of-plane movements, but also because ultrasound energy can implode the bubbles, increasing noise. In addition, it only allows speeds below 0.42 m/s to be measured because it does not have sufficient precision for higher velocities [4].

- **MDCT (Multidetector Computed Tomography)**. Tomography is obtaining images of cuts or sections of a body. The possibility of obtaining images of tomographic sections reconstructed in non-transverse planes has made it preferred to call this technique computed tomography or CT. Instead of obtaining a projection image, like conventional radiography, CT obtains multiple images making the X-ray source and radiation detectors rotate movements around the body. The final representation of the tomographic image is obtained capturing the signals by the detectors and post-processing them using reconstruction algorithms. With this technique, a level of detail less than a millimeter can be obtained for the anatomy over time, making it ideal for studies of cardiac geometry [5]. The term CT often refers to X-ray tomography because it is the most common but there are other types of CT, such as PET (Positron Emission Tomography) and SPECT (Single-Photon Emission Computed Tomography).

- **MRI (Magnetic Resonance Imaging)**. It is a non-invasive technique thanks to which a detailed 3D reconstruction of the anatomy can be performed obtaining images in different measurement planes. Unlike what happens with CT (Axial Computed Tomography), MRI does not use ionizing radiation. It uses magnets that produce a powerful magnetic field that forces the hydrogen atoms in the body to align with that field. When a radio frequency current is pulsed, protons are stimulated and perform an out-of-equilibrium precession motion. When that radio frequency field is turned off, the sensors detect the energy released while the protons are realigning with the magnetic field as well as the time until that realignment. It is necessary to apply post-processing on the images obtained for a proper reconstruction.

Contrast media (often containing gadolinium) is often given intravenously before or during the MRI to increase the rate at which protons realign with the magnetic field, making the image brighter.

PC-MRI (Phase Contrast—Magnetic Resonance Imaging) is a specific type of MRI that measures the velocity of blood flow in any arbitrary direction of the magnetic field gradient, and is based on the principle that the PC-MRI offset is proportional to the speed of the moving proton. Modern PC-MRI is typically time-resolved and provides a flow velocity vector distribution in all three directions of space (usually anterior-posterior, left-right, and superior-inferior). For this reason, it is also known as 3D cine PC-MRI or 4D flow MRI [6].

4.1.2 Acquisition of Images in the Left Atrial Appendage

Due to the complex anatomical characteristics of thrombi, its detection in the LAA, a small area with a multilobed anatomy, can cause difficulties.

- **Transesophageal Echocardiography (2D and 3D TEE).** Imaging with 3D TEE is a relatively recent development that improves the anatomical evaluation of LAA. Although TEE 2D provides higher resolution images, TEE 3D allows a more complete evaluation of LAA overcoming some of the limitations associated with 2D images, such as inadequate image planes. In addition, TEE 3D provides better separation and differentiation between adjacent structures, along with a more complete and comprehensive evaluation of LAA, its complex morphology, and the surrounding structures [7]. The sensitivity of the 3D TEE to detect thrombi in the LAA is still limited, although with recent advances in the development of percutaneous devices for closure of the LAA, 3D TEE has gained in importance in guiding the device to the LAA [8].
- **Intracardiac Echocardiography (ICE).** It can provide multiple views and detailed images of the LAA to diagnose the presence of thrombi [9]. Although ICE is less sensitive compared to TEE for thrombus detection, it can serve as a complementary method, especially when TEE results warrant further evaluation. However, because ICE is an invasive procedure, its use is limited in daily practice and is primarily reserved for catheterization during planned cardiac procedures.
- **Doppler.** Doppler echocardiography is used to assess the risk of thromboembolism in the LAA estimating velocity because TEE has limited sensitivity identifying small thrombi within the lateral lobes, even with 3D images [10].
- **Comparison with Other Imaging Techniques.** Although TEE is the most widely used method for evaluation of the LAA, MDCT and MRI are emerging as valuable modalities for imaging and evaluation of the anatomy and functionality of the LAA. Table 4.1 summarizes the main strengths and limitations of each imaging modality. MDCT and MRI are likely to play an increasingly important role in the preoperative and postoperative assessment of LAA when its image resolution is improved to allow accurate determination of thrombus [10].
  MRI is a non-invasive alternative for those cases where TEE is not possible, such as in patients with esophageal pathologies or who have had a failed probe insertion during TEE. However, this modality has been evaluated in a limited number of studies. MRI can accurately visualize the size and function of the LAA and has the potential to detect thrombi in patients with atrial fibrillation [11]. The sensitivity of MRI identifying the presence of thrombi in the LAA is somewhat lower than with MDCT. Although MRI has many advantages over MDCT and TEE, such as the absence of exposure to iodine contrast and radiation without the need of the introduction of a probe, it still has many limitations for its extended use: low temporal resolution (more than 30 ms), prolonged exploration time, respiratory dependency and impossibility of application in patients with implanted cardiac devices [3]. Furthermore, its spatial resolution, usually greater than 1mm voxel, is not sufficient to address a small region such as the LAA [12].

Table 4.1 Comparison of image acquisition techniques for LAA evaluation [10]

	TEE	MDCT	MRI
Sensitivity/specificity for thrombi detection	92%–100%/ 98%–99%	96%/92%	67%/44%
Spatial resolution	0.2–0.5 5 mm	0.4 4 mm	1–2 mm
Temporal resolution	20–30 ms	70–105 ms	30–50 ms
3D Volume Rendering	Yes (with 3D)	Yes	Yes
Contrast required	No*	Yes	No*
Ionizing radiation	No	Yes	No
Availability	Wide	Limited	Limited
Type of technique	Invasive	Non-invasive	Non-invasive
Real-time evaluation	Yes	No	No
Special considerations		Dynamic evaluation of LAA function	Not applicable to patients with pacemakers

*Contrast can be used to improve visualization of thrombi in doubtful cases

TTE and TEE can be used to measure the velocity of blood flow in both the left atrium and left atrial appendage, but both techniques only provide averaged information of the velocity module and its direction in the measurement volume, without giving detailed 3D information of local flow conditions [13].

4.2   In Vitro Techniques

Over the years, numerous flow visualization techniques have been adapted to study in vitro hemodynamic flows. Flow visualization, combined with advanced data processing techniques, is a powerful tool for deepening qualitative and quantitative knowledge of the fluid field in the heart. The in vitro techniques are optical techniques because they require direct visual access to the study region. These include:

- **PIV (Particle Image Velocimetry)**. Eulerian optical technique that tracks particles in the flow based on cross correlation.
- **PTV (Particle Tracking Velocimetry)**. Lagrangian optical technique that tracks individual tracers in the flow.
- **LDV (Laser Doppler Velocimetry)**. Velocity measurement technique in a measurement volume, determined by the intersection of two coherent light beams.

The Table 4.2 compares the in vitro techniques with the in vivo techniques described previously. From this, it is extracted that PIV is the most used flow diagnostic technique for in vitro studies. It provides higher spatial and temporal resolution than most other techniques, and can be easily adapted to various settings and flow scales.

Table 4.2 Comparison of flow visualization techniques [14]

Technique	Advantages	Disadvantages
PIV	High spatial and temporal resolution	Optical access required
	Potential for volumetric measurements	
PTV	Estimation of Lagrangian statistics: particle trajectories and residence times	Low spatial resolution
		Optical access required
LDV	High spatial and temporal resolution	Point-based measurement
		Optical access required
Color	Non-invasive and in vivo visualization	High velocities limitation (*Aliasing*)
Doppler	Short scanning periods	Semi-quantitative velocity information
Echo-PIV	Non-invasive. It can be used in vivo if the particles resolution is good	Lack of precision for high velocities. Resolution decreases as imaging depth increases
PC-MRI	Non-invasive and in vivo visualization	Spatial resolution lower than PIV
	Potential for volumetric measurements	Expensive and incompatible with metals

It should be mentioned that sometimes it might be necessary to use a combination of flow visualization measurement techniques to get a complete understanding of the flow. For example, a combination of PIV in idealized configurations and Color Doppler in patient-specific geometry models would be a powerful approach to obtain a complete understanding of the fluid field in a clinically relevant environment [14].

4.3 In Silico Techniques

Computational Fluid Dynamics (CFD) is a method widely used in research to simulate blood flow in cardiovascular systems that has experienced remarkable growth within the clinical field in cardiovascular diseases [15–21].

4.3.1 Numerical Resolution of the Navier-Stokes Equations

Fluid flow is governed by the Navier-Stokes equations, but these equations cannot be solved analytically. For this reason, they are numerically solved in a spatial and temporal domain using iterative calculation until the convergence criterion is satisfied, i.e. that residuals are low enough. Every Conservation Equation can be expressed in a generic way:

$$\frac{\partial(\rho\phi)}{\partial t} + div(\rho\mathbf{u}\phi) = div(\Gamma_\phi grad(\phi)) + S \tag{4.2}$$

In Eq. 4.2 ϕ is a specific property, ρ is the density of the fluid, $\mathbf{u}$ is the velocity vector and Γ is the diffusion coefficient of ϕ.

- The first term of Eq. 4.2 corresponds to the **transient** term, which is the temporal variation of the variable ϕ per volume unit.
- The second term represents the **convective** transport, which is the net flow balance of the variable ϕ in a control volume, as a consequence of the velocity field.
- The third term is the **diffusive** transport, which corresponds to the balance of flows of ϕ due to the gradient of ϕ. item The last term, S, is the **source** term. It refers to the net generation of ϕ per volume unit.

There are different cases depending on the property that is replaced by ϕ:

- $\phi = u$: conservation equation of momentum in the x-axis.
- $\phi = v$: conservation equation of momentum in the y-axis.
- $\phi = w$: conservation equation of momentum in the z-axis.
- $\phi = E$: energy conservation equation.

Navier Stokes equations must be transformed to solve numerically:

- It must go from a partial derivative equation to a system of algebraic equations.
- Differential operators must be converted to arithmetic operators.
- The continuous ϕ solution will be a discrete ϕ solution.

For this reason, Eq. 4.2 must be transformed into Eq. 4.3.

$$a_P \phi_P = a_W \phi_W + a_E \phi_E + a_S \phi_S + a_N \phi_N + a_L \phi_L + a_H \phi_H + b \qquad (4.3)$$

The subscripts refer to the faces of the differential cube of Fig. 4.2.

It is necessary to move from a continuous geometric domain to a discrete computational domain. The most widespread discretization methods are the finite difference method, the finite element method and the finite volume method, being the latter the most widely used.

The volume of the study region is divided into small discrete cells that make up a computational mesh, as it is shown in Fig. 4.2.

The conservations equations in each discrete element of the computational mesh are solved to obtain the flow velocity and pressure at each point of the mesh. The computational power required is very high because numerous iterations are necessary until the solution converges. Control of residuals (difference between the result of the previous iteration and the result of the current one) is very important. As these errors decrease, the results of the equation reach values that change less and less, which is known as convergence. On the other hand, if these error start to increase, the solution is said to be divergent.

To solve these equations it is necessary to establish boundary conditions, generally flow rate or pressure at the inputs and outputs of the computational model. Due to the complexity of cardiovascular flows, CFD models often make simplifications (rigid artery walls or consider blood as a Newtonian fluid). These assumptions can affect

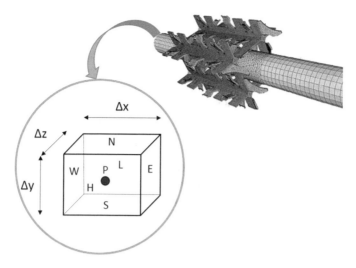

Fig. 4.2 Discretization of the continuous domain in control volumes

the fluid field obtained with CFD in relation to the cardiovascular system. Therefore, the precision of these CFD models must be quantified before being transferred into clinical practice [22].

4.3.2 CFD Model Validation

It is necessary to carry out a validation of the CFD model with experimental measurements that confirm its behavior. Comparative studies of the PC-MRI and PIV measurement techniques have been carried out, which are the most widely used in vivo and in vitro techniques for the validation of CFD models. To quantify the degree of correspondence between the two techniques, the parameter of the root-relative-mean-square error, $RRMSE$, is used, defined as follows:

$$RRMSE = \sqrt{\frac{\sum_t (f_{MRI}(t) - f_{PIV}(t))^2}{\sum_t (f_{PIV}(t))^2}} \quad (4.4)$$

In general, there is a good correspondence between both techniques, although most of the results are limited by PC-MRI resolution. The most relevant studies are detailed below.

Kitajima et al. [23] compared qualitatively PIV and PC-MRI in a patient-specific extracardiac total cavopulmonary connection model. Velocities obtained were of the same order, but they highlighted that the main advantage of PIV over PC-MRI was spatial resolution, which excessively increased the scanning time in PC-MRI.

Hollnagel et al. compared velocity fields obtained with PC-MRA, LDV and CFD in models of cerebral arteries and aneurysms for the case of steady flow [24] and for the case of pulsating flow [25]. The correspondence between LDV and CFD was very high (differences less than 10%), while the comparison with PC-MRA showed differences that can go up to 20%. PC-MRA velocity estimation is best in regions of the straight artery with the measurement plane perpendicular to the direction of the primary flow. Additionally, the 0.49mm x 0.49mm resolution provided by PC-MRA would require a minimum vessel diameter of 2mm in order to obtain detailed and reliable results.

Ooij et al. [26] compared PC-MRI with PIV and CFD. To perform this comparison, they used a life-size intracranial aneurysm model and compared the velocity modulus and the orientation of velocity vectors for steady and pulsating flow. The deviations between PC-MRI and CFD were less than 5% for the velocity modulus in relation to the maximum speed of PC-MRI and at 16° for the angle of the vectors. Regarding the comparison between PC-MRI and PIV, the deviation of the velocity modulus was less than 12% with respect to the maximum velocity of PC-MRI and at 19° for the angle of the vectors.

Khodarahmi et al. [27] compared PC-MRI and stereoscopic PIV and in turn with CFD in models with 50%, 74% and 87% stenosis. They obtained correlation coefficients greater than 0.99 and 0.96 for steady and pulsating flow respectively. The taper errors were less than 5% for both steady and pulsating flow.

Töger et al. [28] studied vortex formation using PC-MRI and PIV. They took the PIV results as a reference and the PC-MRI velocities were consistent with a regression coefficient $R^2 = 0.95$, but the peak velocity obtained was 18% lower than that of the PIV.

Kweon et al. [29] compared PC-MRI with CFD and a flowmeter in models with 75% and 90% stenosis. The flow obtained in the proximal and distal regions of the stenosis by PC-MRI offered good precision measurements with errors of less than 3.6% (resolution of 1.6 6mm) compared to the flowmeter. In contrast, the peak velocity obtained with PC-MRI was 22.8% less than the CFD result.

Puiseux et al. [30] compared PC-MRI and CFD testing pulsating flow in an in vitro model containing an aneurysm, a curved canal and a bifurcation. In the first instance, the velocity correlation between both techniques was poor ($R^2 = 0.63$), but after a post-process in which phase offset errors and velocities near the wall were corrected, they achieved $R^2 = 0.97$ (Table 4.3).

Although in vivo techniques such as PC-MRI reflect better the reality, no pressure boundary conditions can be obtained and the spatio-temporal resolution of the flow velocity measurement in patient's heart may result insufficient in occasions, such as in the LAA.

For this reason, in vitro validation of hemodynamic numerical models, generally using PIV, becomes the most appropriate method [22, 31–33].

In order to perform an adequate validation of the CFD model, it is not enough to only make a qualitative or quantitative comparison, but it is also necessary to establish a methodology for calculating errors such as it is shown in the scheme in Fig. 4.3.

Table 4.3 Summary of studies carried out about the comparison of measurement techniques

Study, year	Application	Techniques	Kind of flow	Conclusions
Kitajima et al. [23]	Extracardiac cavopulmonary connection	PC-MRI PIV	Steady	Qualitative study of velocities. PC-MRI limitation: spatial resolution
Hollnagel et al. [24]	Cerebral artery and aneurysm	PC-MRA LDV	Steady	Low resolution of PC-MRA for vessels with D<2 2 mm
Hollnagel et al. [25]	Cerebral artery and aneurysm	PC-MRA LDV CFD	Pulsating	Deviations between results: PC-MRA/LDV <20% LDV/CFD <10%
Ooij et al. [26]	Brain aneurysm	PC-MRI PIV CFD	Steady and pulsating	Good correlation: PC-MRI/PIV: errors<12% PC-MRI/CFD: errors<5%
Khodarahmi et al. [27]	Stenosis	PC-MRI SPIV CFD	Steady and pulsating	Good correlation (0.99 for steady and 0.96 for pulsating) Errors<5% in the throat
Töger et al. [28]	Vortex	PC-MRI PIV	Steady and pulsating	Good correlation ($R^2 = 0.95$) Peak velocity error: 18%
Kweon et al. [29]	Stenosis	PC-MRI CFD	Steady	Error < 3.6%, except at peak velocity: error >22.8%
Puiseux et al. [30]	Bifurcation and aneurism	PC-MRI CFD	Pulsating	After post-processing $R^2 = 0.97$ (initially $R^2 = 0.63$)

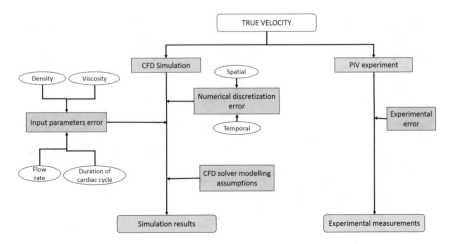

Fig. 4.3 Estimation of errors in CFD models and in experimental measurements using PIV

References

1. Paolinelli GP (2013) Principios físicos e indicaciones clínicas del ultrasonido doppler. Revista Médica Clínica Las Condes 24(1):139–148. Tema central: Radiología al día
2. Otto Catherine M, Schwaegler Rebecca G, Freeman Rosario V (2011) Echocardiography review guide: companion to the textbook of clinical echocardiography: Expert Consult: Online and Print. Saunders
3. Bermejo J, Martínez-Legazpi P, del Álamo JC (2015) The clinical assessment of intraventricular flows. Ann Rev Fluid Mech 47(1):315–342
4. Prinz C, Faludi R, Walker A, Amzulescu M, Gao H, Uejima T, Fraser AG, Voigt JU (2012) Can echocardiographic particle image velocimetry correctly detect motion patterns as they occur in blood inside heart chambers? a validation study using moving phantoms. Cardiovascular Ultrasound 10(1)
5. Lantz J, Gupta V, Henriksson L, Karlsson M, Persson A, Carlhäll C-J, Ebbers T (2018) Impact of pulmonary venous inflow on cardiac flow simulations: comparison with in vivo 4d flow MRI. Ann Biomed Eng 47(2):413–424
6. Itatani K, Miyazaki S, Furusawa T, Numata S, Yamazaki S, Morimoto K, Makino R, Morichi H, Nishino T, Yaku H (2017) New imaging tools in cardiovascular medicine: computational fluid dynamics and 4d flow MRI. Gen Thoracic Cardiovascular Surg 65(11):611–621
7. Nakajima H, Seo Y, Ishizu T, Yamamoto M, Machino T, Harimura Y, Kawamura R, Sekiguchi Y, Tada H, Aonuma K (2010) Analysis of the left atrial appendage by three-dimensional transesophageal echocardiography. Am J Cardiol 106(6):885–892
8. Marek D, Vindis D, Kocianova E (2013) Real time 3-dimensional transesophageal echocardiography is more specific than 2-dimensional TEE in the assessment of left atrial appendage thrombosis. Biomed Papers 157(1):22–26
9. Desimone CV, Asirvatham SJ (2014) ICE imaging of the left atrial appendage. J Cardiovascular Electrophysiol 25(11):1272–1274
10. Beigel R, Wunderlich NC, Ho SY, Arsanjani R, Siegel RJ (2014) The left atrial appendage: anatomy, function, and noninvasive evaluation. JACC: Cardiovascular Imaging 7(12):1251–1265
11. Burrell LD, Horne BD, Anderson JL, Muhlestein JB, Whisenant BK (2013) Usefulness of left atrial appendage volume as a predictor of embolic stroke in patients with atrial fibrillation. Am J Cardiol 112(8):1148–1152

12. Markl M, Lee DC, Furiasse N, Carr M, Foucar C, Ng J, Carr J, Goldberger JJ (2016) Left atrial and left atrial appendage 4d blood flow dynamics in atrial fibrillation. Circulation: Cardiovascular Imaging 9(9)

13. Dentamaro I, Vestito D, Michelotto E, De Santis D, Ostuni V, Cadeddu C, Colonna P (2016) Evaluation of left atrial appendage function and thrombi in patients with atrial fibrillation: from transthoracic to real time 3d transesophageal echocardiography. Int J Cardiovascular Imaging 33(4):491–498

14. Raghav V, Sastry S, Saikrishnan N (2018) Experimental assessment of flow fields associated with heart valve prostheses using particle image velocimetry (piv): Recommendations for best practices. Cardiovascular Eng Technol 9(3):273–287

15. Bosi GM, Cook A, Rai R, Menezes LJ, Schievano S, Torii R, Burriesci G (2018) Computational fluid dynamic analysis of the left atrial appendage to predict thrombosis risk. Frontiers Cardiovascu Medic 5

16. Chung B, Cebral JR (2014) CFD for evaluation and treatment planning of aneurysms: review of proposed clinical uses and their challenges. Ann Biomed Eng 43(1):122–138

17. Dahl SK, Thomassen E, Hellevik LR, Skallerud B (2012) Impact of pulmonary venous locations on the intra-atrial flow and the mitral valve plane velocity profile. Cardiovascular Eng Technol 3(3):269–281

18. Koizumi R, Funamoto K, Hayase T, Kanke Y, Shibata M, Shiraishi Y, Yambe T (2015) Numerical analysis of hemodynamic changes in the left atrium due to atrial fibrillation. J Biomechan 48(3):472–478

19. Olivares AL, Silva E, Nuñez-Garcia M, Butakoff C, Sánchez-Quintana D, Freixa X, Noailly J, de Potter T, Camara O (2017) In silico analysis of haemodynamics in patient-specific left atria with different appendage morphologies. In Pop M, Wright GA (eds.), Functional imaging and modelling of the heart, pp 412–420, Cham, 2017. Springer International Publishing

20. Otani T, Al-Issa A, Pourmorteza A, McVeigh ER, Wada S, Ashikaga H (2016) A computational framework for personalized blood flow analysis in the human left atrium. Ann Biomed Eng 44(11):3284–3294

21. Rayz VL, Boussel L, Acevedo-Bolton G, Martin AJ, Young WL, Lawton MT, Higashida R, Saloner D (2008) Numerical simulations of flow in cerebral aneurysms: comparison of CFD results and in vivo MRI measurements. J Biomech Eng 130(5):051011

22. Paliwal N, Damiano RJ, Varble NA, Tutino VM, Dou Z, Siddiqui AH, Meng H (2017) Methodology for computational fluid dynamic validation for medical use: Application to intracranial aneurysm. J Biomech Eng 139(12):121004

23. Kitajima HD, Sundareswaran KS, Teisseyre TZ, Astary GW, Parks WJ, Skrinjar O, Oshinski JN, Yoganathan AP (2008) Comparison of particle image velocimetry and phase contrast MRI in a patient-specific extracardiac total cavopulmonary connection. J Biomech Eng 130(4):041004

24. Hollnagel DI, Summers PE, Kollias SS, Poulikakos D (2007) Laser doppler velocimetry (LDV) and 3d phase-contrast magnetic resonance angiography (PC-MRA) velocity measurements: Validation in an anatomically accurate cerebral artery aneurysm model with steady flow. J Magn Reson Imaging 26(6):1493–1505

25. Hollnagel DI, Summers PE, Poulikakos D, Kollias SS (2009) Comparative velocity investigations in cerebral arteries and aneurysms: 3d phase-contrast MR angiography, laser doppler velocimetry and computational fluid dynamics. NMR in Biomed 22(8):795–808

26. van Ooij P, Guédon A, Poelma C, Schneiders J, Rutten MCM, Marquering HA, Majoie CB, vanBavel E, Nederveen AJ (2011) Complex flow patterns in a real-size intracranial aneurysm phantom: phase contrast MRI compared with particle image velocimetry and computational fluid dynamics. NMR in Biomed 25(1):14–26

27. Khodarahmi I, Shakeri M, Kotys-Traughber M, Fischer S, Sharp MK, Amini AA (2013) In vitro validation of flow measurement with phase contrast MRI at 3 tesla using stereoscopic particle image velocimetry and stereoscopic particle image velocimetry-based computational fluid dynamics. J Mag Reson Imaging 39(6):1477–1485

28. Töger J, Bidhult S, Revstedt J, Carlsson M, Arheden H, Heiberg E (2015) Independent validation of four-dimensional flow MR velocities and vortex ring volume using particle imaging velocimetry and planar laser-induced fluorescence. Magnet Reson Med 75(3):1064–1075
29. Kweon J, Yang DH, Kim GB, Kim N, Paek MY, Stalder AF, Greiser A, Kim Y-H (2016) Four-dimensional flow MRI for evaluation of post-stenotic turbulent flow in a phantom: comparison with flowmeter and computational fluid dynamics. Europ Radiol 26(10):3588–3597
30. Puiseux T, Sewonu A, Meyrignac O, et al (2019) Reconciling PC-MRI and CFD: an in-vitro study. NMR in Biomed e4063
31. Buchmann NA, Yamamoto M, Jermy M, David T (2010) Particle image velocimetry (PIV) and computational fluid dynamics (CFD) modelling of carotid artery haemodynamics under steady flow: A validation study. J Biomech Sci Eng 5(4):421–436
32. Ford MD, Nikolov HN, Milner JS, Lownie SP, DeMont EM, Kalata W, Loth F, Holdsworth DW, Steinman DA (2008) PIV-measured versus CFD-predicted flow dynamics in anatomically realistic cerebral aneurysm models. J Biomech Eng 130(2):021015
33. Raben JS, Morlacchi S, Burzotta F, Migliavacca F, Vlachos PP (2014) Local blood flow patterns in stented coronary bifurcations: an experimental and numerical study. J Appl Biomater Funct Mater 13(2)

Chapter 5
Techniques for the Validation
of Numerical Models

5.1 MRI

Magnetic Resonance Imaging (MRI) is one of the most used techniques to analyze
the anatomy of the heart, while its PC-MRI variant can obtain a quantification of
cardiac flow.

5.1.1 Operating Principle

In most of MRI equipment, superconducting electromagnets are used. They operate
at the temperature of liquid helium ($-269\,°C$), at which the coils of the conductive
material conduct electricity without resistance and with minimal consumption of
electric current [1].

An MRI machine like the one in Fig. 5.1 has the following elements: magnet, gradient coil, radiofrequency coil, stretcher, where the patient is placed and, sometimes,
an antenna will be necessary, depending on the study zone. Antennas are devices that
serve to detect the signal emitted by protons and adapt to the morphology of the study
region of the body: head, spine, shoulder, knee, or chest. They can be transmitters
(they send the RF pulses), receivers (they pick up the signal emitted by the tissues),
or transmitter–receivers.

Unlike Computerized Tomography, it does not use ionizing radiation, but uses
magnets which produce a powerful magnetic field that forces hydrogen atoms in
substances such as water to align with that field. When the protons are stimulated by
a radiofrequency current, they move out of equilibrium. When this radiofrequency
field is turned off, the protons are realigned with the magnetic field and the energy

---

The original version of this chapter was revised: The author's last name in Refs. 2, 3, 6, 8, 13, 15
of this chapter were incorrect. The author's last names are corrected in References. The correction
to this chapter is available at https://doi.org/10.1007/978-3-030-60389-2_6.

© The Editor(s) (if applicable) and The Author(s), under exclusive license   59
to Springer Nature Switzerland AG 2021, corrected publication 2021
A. Pozo Álvarez, *Fluid Mechanics Applied to Medicine*,
SpringerBriefs in Computational Mechanics,
https://doi.org/10.1007/978-3-030-60389-2_5

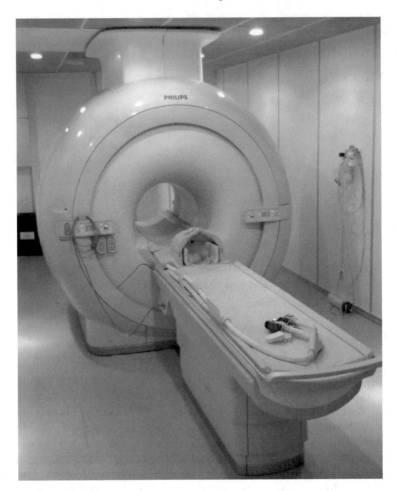

Fig. 5.1 3T MRI Machine

released is measured, as well as the time elapsed until that realignment. These return signals are transformed into images by a computer connected to the scanner.

In the MRI room, the magnet creates such a powerful peripheral field of attraction that it turns any object containing ferromagnetic material into a projectile potential. For this reason, the control of this field is a critical safety factor. The 5-Gauss line (0.5 mT) is established as a safety limit, from which ferromagnetic objects resist the attraction of a magnetic field. For a 3T MRI machine the 0.5mT line is about 5 m from the magnet position. The 0.1 mT line would be about 8 m away.

This will be the biggest restriction of the installation. Aluminum is the only metallic material compatible with MRI. All other metallic materials must remain outside the marked safety limit.

The following stages are distinguished, indicated in Fig. 5.2:

1. Protons present in hydrogen are electrically charged and can be considered as small magnets with their north pole and their south pole. This causes the hydrogen

protons to be susceptible to external magnetic fields. In the absence of an external magnetic field, each proton rotates 360° around its own axis (precession motion) at a certain velocity, the called **Larmor frequency** (Fig. 5.2a. Due to its spin, the proton continuously changes its phase because each phase is instantaneous.

2. However, when hydrogen protons are placed in a strong magnetic field **they are realigned** with the magnetic field. In this resting phase, the net magnetization will always point towards the patient's head, because z-axis represents the magnetic field of the MRI machine (Fig. 5.2b). Despite the parallel alignment of the protons in their resting phase, they do not rotate synchronously (out of phase).

3. Hydrogen protons can be activated by **radiofrequency pulses** with a specific frequency. When the frequency of the hydrogen proton (Larmor frequency) matches the transmitted radiofrequency wave, excitation will occur. All the hydrogen protons will spin spontaneously simultaneously. The transmitted radiofrequency pulse will not only make the protons rotate in phase, but will also rotate their magnetization 90° with respect to z-axis (Fig. 5.2c). The induced magnetic signal changes are recorded by the receiver coils and then processed in the MRI image.

4. **Relaxation**. When the radiofrequency pulse is turned off, protons will return to their original resting phase. Two separate and independent processes occur:

   • Longitudinal relaxation (T1). Protons will return to their original position and the energy received from the radiofrequency pulse will be transferred to their environment (Fig. 5.2d). Relaxation time T1 is defined as the time required to reach 63% of the original longitudinal magnetization (in z-axis). Each tissue has its own time and T1 relaxation curve. The blue curve is for water while the yellow is for adipose tissue.

   • Transverse relaxation (T2). Simultaneously, protons that rotated in phase will no longer rotate in sync once the radiofrequency wave has been turned off (out of phase). This process is called lag and occurs because the magnetic field of the MRI machine is no longer completely homogeneous. The net magnetization consists of a longitudinal component (z-axis) and a transverse component (xy axis). Together they constitute the net magnetization vector. Relaxation T2 describes what happens in the xy plane (Fig. 5.2e).

It is common that intravenous contrast media (often gadolinium) is administered intravenously before or during the MRI to increase the rate at which protons realign with the magnetic field (shorten relaxation time T1). This causes that resulting images are brighter.

The gradient coil is available to determine the location of the protons in each axis. It has three magnets (called x, y, z) less powerful than the main magnet, each of them oriented along a different plane. They modify the magnetic field at specific points and work together with radiofrequency pulses to generate the scanner image encoding the spatial distribution of the protons. When turned on and off quickly, gradients allow the scanner to display the body image in slices.

However, the x, y, z gradients can be used in combination to generate image cuts in any direction, which is one of the great strengths of MRI as a diagnostic tool.

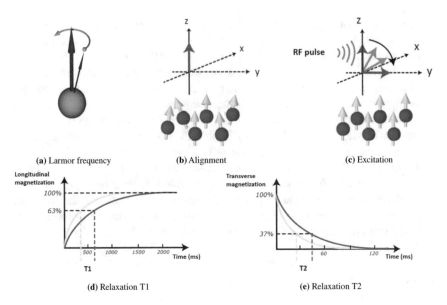

Fig. 5.2 Proton movement during MRI

Thorax is studied using strict axial (xy), coronal (xz), and sagittal (yz) planes, but the cardiac axes are not parallel to the body axes (the long axis of the heart is located around 45° respect to the midsagittal plane of the spine dorsal). Therefore, the heart must be studied using specific planes. Initially, single-shot sequences are used to locate the heart, in which spatial and temporal resolution is sacrificed to obtain images quickly [2].

The intrinsic axes of the heart that should be studied in an MRI scan are: four chambers, two chambers (long axis of the left ventricle), short axis, three chambers (exit tract of the left ventricle), and exit tract of the right ventricle [2] (Fig. 5.3).

In order to obtain clear images of the heart, it is necessary to introduce systems that minimize or eliminate the effect of the physiological movements of the heart and respiration (the least predictable). In addition, nonmagnetic electrodes must be placed on the patient's chest to achieve cardiac synchronism.

5.1.2 Image Registration Using MRI

An MRI sequence generally consists of several radiofrequency pulses. The difference between the sequences is given by the type of radiofrequency pulses used and the time between them. The MRI receiver catches the changes in the magnetization that occur in protons after having received the energy of the radiofrequency pulses (the longitudinal and transverse relaxation of the protons). The information obtained is

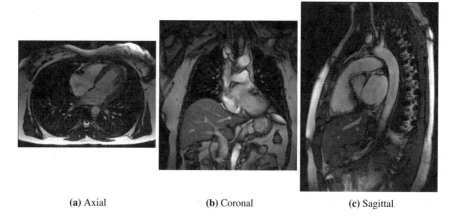

(a) Axial   (b) Coronal   (c) Sagittal

Fig. 5.3 Orthogonal planes of the heart obtained with MIR

Table 5.1 MRI imaging enhancement

	T1-weighted	T2-weighted	PD-weighted
Hyperintense	Short T1	Long T2	More protons density
Hypointense	Long T1	Short T2	Less protons density
TR	Short (<500 ms)	Long (>2000 ms)	Long (2000–2500 ms)
TE	Short (<25 ms)	Long (80–120 ms)	Short (<25 ms)

called the raw data matrix or space k. This is a grid of points to which the Fourier transform must be applied to obtain an interpretable image [1] (Table 5.1).

The following parameters must be adjusted for image acquisition:

- Echo Time (TE). Time between the application of the radiofrequency pulse and the peak of the signal induced in the coil. It controls the relaxation time T2.
- Repetition Time (TR). Time from the application of the excitation pulse to application of the next pulse. It determines the longitudinal magnetization that is recovered between each pulse.
- A contrast medium (gadolinium) can be added intravenously to increase the proton's velocity during the realignment with the magnetic field (shorten the T1 relaxation time).

Sequences are obtained combining these variables. The result can be to differentiate tissues according to their T1 (T1-weighted images) or according to their T2 (T2-weighted images). There are also sequences that enhance the image with high proton density (PD) [3].

Water has a long T1 (3000 ms) and a long T2 (3000 ms), so it is hypointense in T1 and hyperintense in T2. In contrast, a mixture of distilled water (61% by weight) with glycerin (39% by weight) would have a T1 of approximately 500 ms and a T2 of approximately 45 ms [4].

The sequences that are commonly used in clinical practice in an MRI study are divided into pulse sequences (anatomical), gradient-echo sequences (anatomic and cine), flow sequences, and 3D gradient-echo sequences [2].

5.1.2.1 Spin Echo sequences

Spin Echo sequences (SE) are characterized because the blood inside the vessels and the cardiac chambers is hypointense or black. These sequences are obtained with selective RF pulses for each slice, so that blood flowing through the plane during MRI imaging will not provide signal (intravascular signal void). Modifying the acquisition parameters, images enhanced in T1, proton density, or T2 can be obtained. T1-weighted SE sequences (SE-T1) are used to obtain anatomical information. These sequences are obtained with electrocardiographic synchronization, and in them the repetition time (TR) must be equal to the RR interval of the patient's ECG.

Since the enhancement of the images depends on the TR and the TE, when the TR is equal to the time of an RR, the TE must be short in order to obtain SE-T1 images, while when the TR is equal to 2 times or more RR, TE has to be long to get SE-T2 images. The number of cuts of these sequences depends on the TR and, therefore, the patient's heart rate.

Data required to create the image for each section is acquired at the same phase of the cardiac cycle, so that section can be obtained at various instants in the cardiac cycle.

Modifications have been made from conventional SE sequences that allow obtaining images much more quickly. These black blood sequences have been called Turbo Spin Echo (TSE), Fast Spin Echo (FSE), or double inversion pulse sequences [1].

5.1.2.2 Gradient Echo Sequences

Gradient Echo sequence (GE) provides images with white blood. The hypersignal of the blood is due to its movement, which contrasts with the signal loss due to saturation of the tissues. It will be more hyperintense when the flow direction is perpendicular to the image plane. GE sequences are used to study the functionality of the heart. Its main characteristic is the high temporal resolution (it allows acquiring an image at 20–40 ms intervals during the cardiac cycle), which allows it to be analyzed in cine mode. These sequences can be obtained in a single section (a multiphase cut) or in multiple sections (multi-cut-multiphase).

Following conventional GE sequences, a second generation of Fast-GE sequences has been developed, known among a multitude of dependent acronyms from different commercial houses, such as Turbo-GE and Fast-GE. The most recent sequences used in clinical practice are capable of obtaining a large number of images in a few ms, they are called Balanced Fast Field Echo (Philips), TrueFISP (Siemens) and Fiesta (Fast Imaging Employing Steady-state Acquisition) [1].

5.1.2.3   Phase Contrast Sequences

GE sequences with phase contrast allow visualization and quantification of blood flow. This quantification can be carried out in the plane of the flow direction and/or in a plane perpendicular to the flow. They are based on the fact that protons that move along a magnetic field change the direction of the phase in proportion to the velocity and intensity of the gradient. These sequences provide information about the magnitude and phase of the flow. In magnitude images the flow is hyperintense, while in phase images the flow can be hyperintense or hypointense, depending on its direction. In the phase images, velocity/time or flow/time curves can be obtained in order to allow quantifying the velocity of the flow and the pressure gradients in the vessels.

5.1.2.4   3D MRI Angiography

T1-weighted 3D Gradient Echo sequences after contrast administration (Gadolinium, Gd-DTPA) are used for MRI 3D Angiography studies. Volumetric images synchronized with intravenous injection of contrast material are acquired, with high resolution and wide fields of vision during apnea, without the need of electrocardiographic synchronization. This sequence is useful to study smaller blood vessels.

5.1.3 Resolution

The MRI used in medicine can reach resolutions of up to 1 mm.

When a body is placed in a magnetic field B_0, the magnetic moment of the protons in the nucleus of the hydrogen atoms in the water precedes a frequency that depends on the strength of the magnetic field according to the Larmor Equation [5]:

$$f_0 = \frac{\gamma B_0}{2\pi} \tag{5.1}$$

In Eq. 5.1, γ is the gyromagnetic relation. For hydrogen $\frac{\gamma}{2\pi} = 42.57\,57\,\text{MHz/T}$ and f_0 is in the 50–500 500 MHz range for protons. The magnetic moments are aligned in a direction parallel to the main magnetic field, establishing a net magnetization in the tissue water. A radiofrequency (RF) coil produces a magnetic field B_1, which changes the direction of magnetization with the sequence of pulses (for example, 90° and 180°). Following RF pulses, the radiofrequency coil detects the return of the magnetization to equilibrium. Spatial localization is generated with the use of gradient coils (G_x, G_y, G_z). Variation of the Repetition Time (TR) and echo time (TE) in MRI pulse sequence provides the basis for different contrast mechanisms (T1- or T2-weighted images). MRI signal is acquired as a complex variable (magnitude and phase). The magnitude is sufficient to represent anatomical features, measure relaxation times, or estimate the diffusion coefficient of water, while phase is used to measure blood flow, temperature, and tissue stiffness.

For a Spin Echo sequence, the signal intensity is proportional to the local concentration of water multiplied by the expression of exponential damping $(1 - e^{-TR/T1})e^{-TE/T2}$. So changes in the tissues will affect the amount of water or change T1 or T2. For fixed values of TR and TE, they will alter the magnitude and phase of the detected signal.

In an MRI image, the Field Of View (FOV) for a single plane can be decomposed into individual image elements (pixels) that determine the resolution of the image. The ratio between pixels and FOV in the x and y directions is

$$FOV_x = N_x \cdot \Delta x \qquad FOV_y = N_y \cdot \Delta y \tag{5.2}$$

In the expressions above, N_x and N_y are typically 128 or 256.

MRI acquisition involves sampling the detected signal for every plane of the object to be studied. The sampling frequency must be less than twice the maximum frequency (Nyquist criterion). This constraint provides the following MRI image resolution equations for a spin echo pulse sequence [6].

$$\Delta x = \frac{FOV_x}{N_x} = \frac{2\pi}{N_x \gamma G_x T_{acq}} \qquad \Delta y = \frac{FOV_y}{N_y} = \frac{2\pi}{N_y \gamma G_y T_{pe}} \tag{5.3}$$

G_x and G_y correspond to the magnitude of the encoding gradients x (read) and y (phase) (typically 1–200 G/cm). T_{acq} is the duration of the acquisition window (typically 1–10 ms) and T_{pe} the phase encoding gradient (usually less than 5 ms). From Eq. 5.3, it follows that the stronger gradients (G_x, G_y, G_z) improve the resolution. The location and thickness of the plane are determined modulating the transmitted radiofrequency pulses (B_1).

The spatial resolution in the z-direction can be obtained using Eq. 5.4.

$$\Delta z = \frac{4\pi}{\gamma G_z \tau_p} \tag{5.4}$$

Signal-to-Noise Ratio (SNR) associated with MRI can be expressed by Eq. 5.5.

$$SNR \propto B_0^2 V_s N \tag{5.5}$$

B_0 is the static magnetic field, V_s is the sample volume, and N is the number of images acquired. MRI machines of 3 3 T or more directly improve SNR and allow faster images to be obtained without compromising resolution. The ability to distinguish individual pixels in an MRI image is known as contrast. Image noise often limits the achievable contrast, so contrast-to-noise ratio, CNR, defined for two tissue regions A and B, is used as an indicator [6]:

$$CNR = \frac{SNR_A - SNR_B}{Noise} \tag{5.6}$$

Contrast improves if the difference between the signal strengths of two adjacent regions of an image is increased and the noise level is reduced, which can be estimated from Eq. 5.7, where K_B is the Boltzman constant, T is the temperature, Δf is the frequency bandwidth of the detector, and R is the general resistance of the loaded RF coil.

$$Noise = \sqrt{4K_B T \Delta f R} \tag{5.7}$$

5.2 PC-MRI

This technique is also called 4D flow MRI or cine because it permits to obtain the temporal variation of the flow in the three dimensions of space for a specific study region.

5.2.1 Operating Principle

Its operating principle is based on variations in the Larmor precession frequency, ω_L, under a magnetic field as it is detailed below [7].

$$\omega_L(\mathbf{r}, t) = \gamma B_0 + \gamma \Delta B_0 + \gamma \mathbf{r}(t)\mathbf{G}(t) \tag{5.8}$$

where γ is the gyromagnetic relation, $\mathbf{r}$ is the displacement, B_0 the static magnetic field, ΔB_0 is the lack of homogeneity of the local field, and $\mathbf{G}$ is the gradient of the magnetic field.

Assuming that the fluid velocity is constant during acquisition, the displacement can be expressed as

$$\mathbf{r}(t) = \mathbf{r_0} + \mathbf{v}(t - t_0) \tag{5.9}$$

In the previous expression, t_0 is the acquisition time and $\mathbf{r_0}$ the offset for t_0.

The phase change of the fluid with the velocity $\mathbf{v}$ under the magnetic gradient is obtained integrating ω_L from t_0 to the echo time (TE):

$$\phi(\mathbf{r}, TE) = \phi_0 + \gamma \mathbf{r_0} \int_{t_0}^{TE} \mathbf{G}(t)dt + \gamma \mathbf{v} \int_{t_0}^{TE} \mathbf{G}(t)t dt + \cdots = \phi_0 + \gamma \mathbf{r_0} M_0 + \gamma \mathbf{v} M_1 + \cdots \tag{5.10}$$

The first term of Eq. 5.10 is the background phase offset ϕ_0, caused by the lack of homogeneity of the field. The second and the third describe the influence of the magnetic gradient $\mathbf{G}$ on the stationary (M_0) and moving (M_1) spins, respectively.

Conventional 2D PC-MRI uses a bipolar gradient along the flow-encoding direction before the reading sequence. The bipolar gradient removes phase accumulations from stationary spins, causing only the first and third term of Eq. 5.10 to take effect. Furthermore, a PC-MRI sequence normally makes two acquisitions with different values of M_1 to remove the unknown phase offset ϕ_0.

There are two acquisition methods for 2D PC-MRI. The first uses the flow compensation gradient with the flow-encoding gradient to obtain the reference image. The second employs the combination of the flow-encoding gradient and the following gradient with the opposite polarity. In 2D PC-MRI, the bipolar gradient is mainly used, while in 4D PC-MRI the first acquisition method is the most widely used [7]. Figure 5.4 shows a scheme of these two methods.

Phase offset difference between two scans is directly related to the flow velocity using the following expression:

$$\Delta\phi = \gamma \mathbf{v} \Delta M_1 \tag{5.11}$$

2D PC-MRI uses two acquisitions, while 4D PC-MRI typically uses four scan points measuring velocity encoding in the three directions of space and flow compensation encoding, as it is shown in Fig. 5.4b. Assuming that the background phase

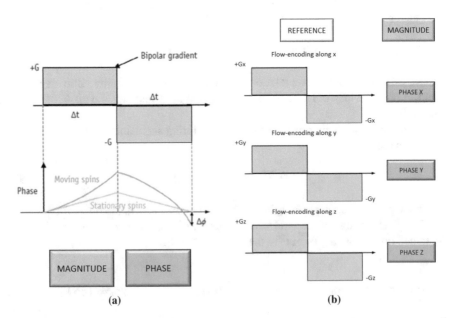

Fig. 5.4 Principles of 2D PC-MRI (**a**) and 4D PC-MRI (**b**)

Table 5.2 Procedure for using 4D PC-MRI to measure hemodynamic flows

MRI scanner	External processing	
Scan parameters	Preprocessing	Data analysis
Field of View	Phase offset corrections	Velocity field
Spatial resolution	Phase aliasing corrections	Streamlines
Temporal resolution	Mask generation	Flow rate
Velocity encoding		Wall shear stress
K-space segmentation		Turbulent kinetic energy
Flip angle		Vorticity
ECG synchronization		Relative pressure drop

offset is the same for all scans, the velocity in the three directions can be obtained replacing the phase shifts in the three-way encodings with the phase in the reference image.

5.2.2 Scan Parameters

When 4D PC-MRI is used to acquire images, it is necessary to adjust a series of parameters to obtain images of enough quality in a reasonable scan time (Table 5.2).

Dyverfeldt et al. [8] performed an analysis of the PC-MRI scanning parameters, being the most important the field of view, spatial resolution, temporal resolution, and velocity V_{enc}.

- A very large **Field Of View** (FOV) requires more data to scan, and therefore more scan time. So the FOV should correspond to the smallest region that contains the region of interest to reduce scanning time.
- The parameter of **velocity encoding**, V_{enc} determines the maximum velocity based on the difference of the first gradient momentum ΔM_1:

$$V_{enc} = \frac{\pi}{\gamma \Delta M_1} \rightarrow v = V_{enc} \cdot \frac{\Delta \phi}{\pi} \tag{5.12}$$

If the detected velocity coincides with the maximum, it will cause a phase offset of π value. On the other hand, if the flow velocity is higher than the V_{enc} it will cause aliasing in the phase, so the V_{enc} must be higher than the expected maximum velocity of the flow. However, the higher the V_{enc} the lower the Signal-to-Noise Ratio (SNR), so a V_{enc} 10% higher than the expected maximum speed is recommended [8].

- The **spatial resolution** should be as high as possible to achieve more accurate results in terms of flow quantification, but the smaller the voxel, the greater the scan time and the lower the SNR. So a compromise solution must be established between spatial resolution, scan time, and SNR. In practice, voxel sizes of 2.5–3 3 mm and 0.7–1.5 5 mm are the most used to measure cardiac and intracranial blood flows, respectively [8].
- The **temporal resolution** should be as small as possible to adequately identify temporal variations in pulsating blood flow. However, in 4D PC-MRI, the number of acquisitions made with MRI is directly proportional to the number of time steps required to describe the entire pulsating cycle. Therefore, the temporal resolution is reduced so as not to excessively increase the scanning time. The usual temporal resolution is around 40 ms.
- The **ECG synchronization** should be retrospective if it is available. Despite the complexity of the reconstruction, the aim is to cover the entire cycle and avoid interruption of the sequence.
- The ideal **segmentation factor k** would take the value of 1, which would increase the temporal resolution but in turn would increase the scanning time, so the value of 2 is usually used.
- A full **coverage of the k-space** is sought in the phase and cut encoding directions to increase the Signal-to-Noise Ratio (SNR) and resolution but the limiting factor is the scan time. If it is possible, the best solution would be an elliptical k-space.
- The **flip angle**. The Ernst angle, α, is the flip angle that maximizes the signal in T1-weighted sequences with a short repetition time (TR).

$$\alpha = acos(e^{-TR/T1}) \tag{5.13}$$

However, depending on the situation, a less than optimal flip angle can be chosen to sacrifice the maximum signal for better T1 contrast.

In T2-weighted sequences, a nonoptimal flip angle can be deliberately chosen to decrease the T1 signal.

- The **movement caused by breathing** must be compensated to try to achieve 100% acceptance. The best solution to reduce scanning time and respiratory artifacts (phantom and blur) is to use the initial or final navigator on the liver/diaphragm interface and a window size of 6 6 mm, which generally results in 50% of acceptance. If this method is not possible, bellows with 50% acceptance could be used.
- **Eddy currents** originate from the variation of the magnetic field, which induces eddy currents in the surrounding conductive materials according to Faraday's Law of Induction. There are mainly two ways to compensate eddy currents: pre-distort gradient pulse, so the magnetic field generated by eddy currents improves the original magnetic field, or apply a secondary coil around the main gradient coil to reduce or cancel eddy currents generated.

5.3 PIV

The Particle Image Velocimetry technique, known as PIV, is the most widely used technique for in vitro validation. It allows determining the velocity field registering the displacement of thousands of particles simultaneously. PIV is a technique that allows visualization of flow as well as quantitative velocity measurements.

5.3.1 Operating Principle

The procedure is summarized as follows: a series of particles that follow their movement called tracer particles are added to the fluid. A digital camera, perfectly synchronized with a laser source, collects pairs of consecutive pictures of a test area recording the position of the illuminated particles. A diagram of an installation that allows measuring using the PIV technique is shown in Fig. 5.5.

If laser is pulsed twice with a known time interval, two positions of every particle will be recorded, thus knowing the displacement suffered by a said particle during that time interval. Subsequently, the velocity of every particle is deduced as the quotient of the displacement and the time interval. Images are analyzed using specific software.

Particle Image Velocimetry is based on the measurement of the velocity of the tracer particles added to the fluid. For this, the plane to be investigated is illuminated with a laser light sheet, so particles are visualized and their image can be registered. The analysis is performed with two consecutive images of the same region at two instants t and t', separated by a time interval Δt. The images are divided into grids, called interrogation areas, so everyone is analyzed separately. In each interrogation area, the two images are superimposed, the coordinates of each point of light are

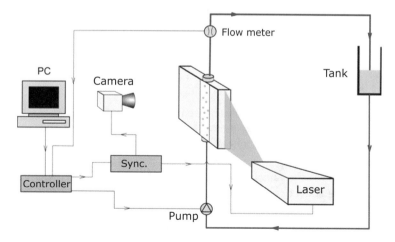

Fig. 5.5 Diagram of a experimental rig to measure with PIV technique

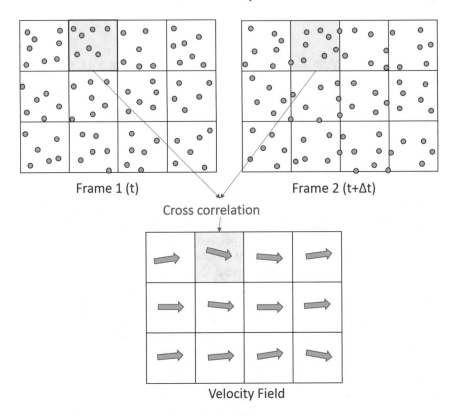

Fig. 5.6 Process of obtaining velocity field using PIV

calculated and the difference between all the points is measured to determine the particles' displacement $\Delta\mathbf{x}$. Making a correlation of all these vectors, the instantaneous velocity vector $\mathbf{U}(\mathbf{X}, t)$ of the fluid is obtained in that interrogation area (Fig. 5.6):

$$\mathbf{U}(\mathbf{x}, t) = \frac{\Delta\mathbf{x}(\mathbf{x}, t)}{\Delta t} \tag{5.14}$$

The greater the number of particles in each interrogation area, the greater the precision achieved with each velocity vector. And with a higher density of interrogation areas, the number of vectors increases, and therefore the velocity field will have a higher resolution.

Figure 5.7 shows the velocity field obtained using a 2D-2C PIV system. It is a model of a straight and rigid duct in which a bluff body with a square section of 0.5 5 mm side has been embedded, located approximately on the axis of symmetry. The incident flow in the obstacle is laminar and it can be seen the wake downstream of the bluff body. $\|V\|$ refers to the velocity module (Fig. 5.8).

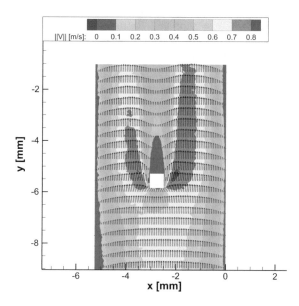

Fig. 5.7 Velocity field obtained with a 2D-2C PIV system

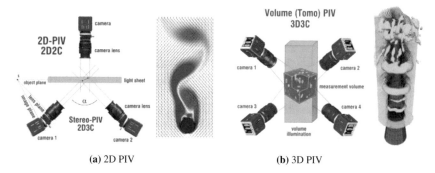

(a) 2D PIV   **(b)** 3D PIV

Fig. 5.8 Types of PIV systems (Adapted from: http://lavision.de/de/applications/fluid-mechanics/index.php)

There are several spatial configurations for PIV systems, and the following classification can be established:

- **Plane (2D-2C)**. 2D system that can only capture two velocity components. It has a camera and a light source of at least 50 mJ/pulse [9].
- **Stereoscopic (2D-3C)**. 2D system that can capture three velocity components. It has two cameras and a light source with at least 50 mJ/pulse [9] is required. To acquire the third velocity component, the second camera is directed at the illumination plane from a different angle allowing reconstruction of the velocity component out of the plane. Measuring on multiple planes the 3D fluid field can be reconstructed interpolation during post-processing [10].

- **Volumetric (3D-3C)**. 3D system that captures three velocity components. It has multiple cameras, the number varies according to the technique (Tomographic, Holographic, Defocusing, Phenoptic). High-power light source (over 200mJ/pulse) is required [9]. One of the most appropriate techniques for acquiring complex 3D flows is TomoPIV, which usually uses four cameras that take simultaneous images from various directions to reconstruct the 3D velocity field with a 3D correlation [11]. This method supposes a great initial investment and the computational requirements are high, especially the processing time [12].

5.3.2 PIV Elements

The main elements of a PIV setup are the laser, the digital camera, and the synchronizer, which is a device used to control the camera and the laser. Your goal is that shots from the laser device and camera captures occur synchronously. In addition, it must be compatible with these devices and the control software.

However, more attention will be paid here to the choice of fluid and seeding particles to properly validate hemodynamic numerical models.

5.3.2.1 Model

Figure 5.9 shows the CAD model of a coronary artery bifurcation in which a stent has been implanted in its main duct. The secondary duct is placed 45° from the main duct.

The first step is to choose the material of the model to test, which must be optically accessible. Models can be made of silicone but one of the most used materials is **methacrylate** (PMMA). The model in the latter case would be rigid because the modulus of elasticity of the methacrylate is 3300 MPa.

Table 5.3 shows the refractive indices of some transparent materials.

5.3.2.2 Fluid

The main element to be defined is the fluid. It must be transparent because optical access is required with the PIV technique. That is why it is important that the **refractive indexes of the fluid and the model are compatible** to obtain images that reflect as closely as possible the real image inside the model. If what is obtained are highly distorted images, the work to correct the images' deformation is highly laborious, especially in the area near the duct walls.

Wright et al. [13] performed a review of the fluids used for measurement techniques that require optical access, such as PIV. One of the most widely used is a mixture of distilled water and glycerin.

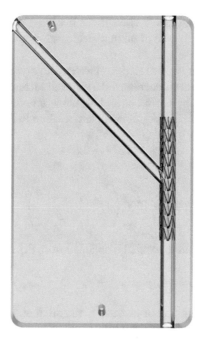

Fig. 5.9 CAD model of a coronary artery bifurcation in which a stent has been implanted

Table 5.3 Refractive indices of some transparent solid materials

Material	Refractive index
FEP	1.34
Silicone elastomer	From 1.4 to 1.44
Fused quartz	1.46
PVA	1.47
VeroClear	1.47
Silica gel	1.47
Glass (Pyrex)	From 1.47 to 1.49
Methacrylate (PMMA)	1.49
UOPTIC2	1.5
Polycarbonate	1.58

Mixture of distilled water and glycerin

The proposed fluid would be a mixture of distilled water at 39% by weight and glycerin at 61%. The resulting mixture has a refractive index of 1.41, fully compatible with the refractive index of the model, which would be made of methacrylate, which takes a value of 1.49.

The mixture has a density of 1158 kg/m^3 and a dynamic viscosity of 0.0094 Pa s for a temperature of 25 °C. The value of these properties changes with temperature, especially in case of the dynamic viscosity as it will be seen later. As an example, the refractive index for this type of mixture decreases around 0.04% each increment of 5 °C.

The value of the density of the mixture of water with glycerin does not have a great variation with temperature (it decreases 0.21% each increment of 5 °C). Equations 5.15 and 5.16 proposed by Cheng et al. [14] show the variation of the density of water ρ_w and of glycerin ρ_{gl} in kg/m^3 with temperature in °C.

$$\rho_{gl} = 1277 - 0.654 \cdot T \tag{5.15}$$

$$\rho_w = 1000 \left(1 - \left| \frac{T-4}{622} \right|^{1.7} \right) \tag{5.16}$$

The density of the mixture would be obtained from Eq. 5.17.

$$\rho_m = \rho_{gl} \cdot C_{Wgl} + \rho_w \cdot (1 - C_{Wgl}) \tag{5.17}$$

C_{Wgl} is the mass concentration of glycerin in the mixture, in this case 0.61.

The dynamic viscosity of the mixture of distilled water and glycerin will be calculated from the temperature and the mass ratio of water–glycerin. The viscosity of the mixture follows Eq. 5.18 proposed by Cheng [14].

$$\mu_m = \mu_w^\alpha \cdot \mu_{gl}^{1-\alpha} \tag{5.18}$$

In Eq. 5.18, μ_m is the viscosity of the mixture, μ_w is the viscosity of water, and μ_{gl} is the viscosity of glycerin, all of them in Pa s. α is a parameter that depends on the temperature T in °C and the mass concentration of glycerin C_{Wgl}.

$$\alpha = 1 - C_{Wgl} + \frac{a \cdot b \cdot C_{Wgl}(1 - C_{Wgl})}{a \cdot C_{Wgl} + b(1 - C_{Wgl})} \tag{5.19}$$

$$a = 0.705 - 0.0017 \cdot T \qquad b = (4.9 + 0.036 \cdot T)a^{2.5}$$

$$\mu_w = 0.00179 \cdot e^{-\left[\frac{(1230+T)T}{36100+360T} \right]} \tag{5.20}$$

$$\mu_{gl} = 12.1 \cdot e^{-\left[\frac{(1233-T)T}{9900+70T} \right]} \tag{5.21}$$

In Fig. 5.10, it can be seen how viscosity is the most sensitive property to changes in temperature. The dynamic viscosity of the mixture decreases its value by 23% for each temperature increase of 5 °C due to the viscosity of glycerin.

It is worth noting the importance of controlling the temperature, since changes in the dynamic viscosity notably influence the Reynolds number, which increases approximately 4.5% for each temperature unit increase, as it can be seen in Fig. 5.10.

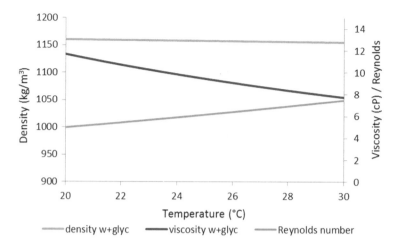

Fig. 5.10 Properties of the mixture of water and glycerin as a function of temperature

The blood density is close to 1055 kg/m³ and shows a non-Newtonian behavior. For the study of most cases, the simplification of assuming that blood behaves like a Newtonian fluid with an average viscosity of 0.0035 Pa s is considered acceptable because the difference in the results would be small and savings in complexity is considerable when taking measurements.

On the other hand, the working temperature does not have to coincide with the blood temperature in the human body, but it must remain constant because experiments do not require changes in viscosity, density, or Reynolds number, which would be a consequence of a change in temperature.

The criteria of transparency and refractive index are adequately matched, but it is not the same with density and viscosity. The density is 9% higher than that of blood, while the parameter that is worst adjusted is the dynamic viscosity, since the viscosity of the mixture is about three times higher than the average viscosity of blood.

Despite being a significantly high difference, a priori the viscosity of the mixture of water and glycerin would not be a problem because what really must be matched between the actual flow to reproduce and that reproduced in the PIV setup is **the Reynolds number**. If the flow is steady it is enough with the coincidence of Reynolds numbers, but if it is pulsating Womersley numbers must match too. Due to the high dynamic viscosity of the mixture, it may be necessary to reduce it considerably precisely in order to achieve the Reynolds numbers that are reached in the heart. For this, another fluid could be used such as distilled water with dissolved Sodium Iodide (NaI), proposed below.

Sodium Iodide solution
Varying the concentration of Sodium Iodide, the refractive index of the mixture can be adjusted to match that of the model. As a limit is the solubility of Sodium Iodide in water, which increases with temperature according to Table 5.4.

Table 5.4 Solubility of NaI in water for different temperatures [15]

Temperature (°C)	23	30	35	40
Solubility (%)	65	66.1	66.8	67.5
Refractive index	1.499	1.503	1.507	1.510

The refractive index of the fluid increases with the concentration of Sodium Iodide present in the water. Dependency is quadratic as it can be seen in Eq. 5.22 proposed by Bai et al. [15].

$$n = 0.2425c^2 + 0.09511c + 1.335 \tag{5.22}$$

To match the refractive index of methacrylate (1.49), the necessary concentration by weight of NaI (c) is 61%.

Kunlun and Joesph [15] proposed Eq. 5.23 to calculate the density in solutions of NaI with mass concentrations of NaI between 10 and 63% and temperatures between 10 and 90 °C.

$$\log_{10} \rho = \log_{10} \rho_0 + (b_0 + b_1 T + b_2 T^2)c \tag{5.23}$$

In the previous expression, the density ρ is in kg/m^3 and T in °C. The value of the constants is $b_0 = 4036.7 \times 10^{-4}$, $b_1 = -294.9 \times 10^{-6}$ and $b_2 = 183.2 \times 10^{-8}$. The density of pure water, ρ_0, can be estimated from Eq. 5.24.

$$\rho_0 = 1000 - 0.062T - 0.00355T^2 \tag{5.24}$$

As with the mixture of water and glycerin, the value of the density of the NaI solution remains practically constant in the range of working temperatures, it only decreases 0.26% for each increment of 5 °C. Figure 5.10 illustrates the variation of the density of the mixture of water and glycerin and of the NaI solution with temperature.

In the case of NaI solution, the variation of dynamic viscosity with temperature is less than if the mixture is of water and glycerin. The dynamic viscosity of the NaI solution can be estimated from Eq. 5.25 proposed by Bai et al. [15] for the range of NaI mass concentrations from 10 to 63% and temperatures from 10 to 90 °C.

$$\log_{10} \mu = \log_{10} \mu_0 + (d_0 + d_1 T)c \tag{5.25}$$

In the previous expression, the viscosity μ is in kg/(m s) and T in °C. The value of the constants is $d_0 = 64 \times 10^{-2}$ and $d_1 = -55.8 \times 10^{-4}$. c is the mass concentration of NaI and μ_0 corresponds to the viscosity of pure water, which can be estimated from Eq. 5.26.

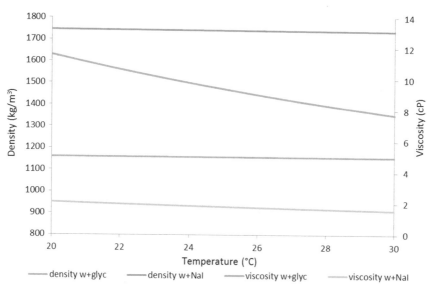

Fig. 5.11 Densities and viscosities of the proposed fluids

$$\mu_0 = 0.59849(43.252 + T)^{-1.5423} \tag{5.26}$$

The viscosity of the NaI solution decreases with temperature of approximately 14% for each temperature increase of 5 °C.

As it is shown in Fig. 5.11, for a temperature of 25 °C, the density of NaI dissolved in water and the resulting dynamic viscosity are, respectively, 1741 kg/m^3 and 0.0018 Pa s.

As the density of the NaI solution is higher and the viscosity is about 5 times lower than that of the mixture of water with glycerin, higher Reynolds numbers can be achieved.

However, the main drawback derived from its use is that after a few hours of its preparation it begins to gradually change color from yellow to red due to oxidation of Sodium Iodide. Adding about 0.5 5 g of sodium thiosulfate in a 1 1 L solution, the fluid will temporarily become transparent again [16].

5.3.2.3 Seeding Particles

They are small solid particles that accompany the fluid in its movement so that its velocity is representative of that of the fluid without altering its properties. Their position should be easily recorded when they are illuminated. The requirements for the seeding particles are

1. **They must follow adequately the flow**. This requirement is related to the density of the material of the particle and its size. These particles are theoretically considered as spherical elements that move within a fluid. Using too small tracer particles will cause a relative error due to Brownian motion [17]. To study this error, the colloidal diffusion coefficient D_i is defined, based on the relation between thermal agitation and viscous forces.

$$D_i = \frac{K \cdot T}{3\pi \mu d_p} \tag{5.27}$$

In Eq. 5.27, μ is the viscosity of the fluid, T is the temperature, d_p the diameter of the particle, and K the Boltzmann constant. The larger the diameter of the particles, the less important Brownian motion will be. The standard deviation of a random displacement of a particle is given by Eq. 5.28.

$$\sigma_p = \Delta x_p \approx \sqrt{2 D_i \delta t} \tag{5.28}$$

The random displacement error will increase if the particles are too small. But the diameter of the particles cannot be increased excessively because it would increase the probability of sedimentation in the flow. For this reason, an intermediate solution must be found.

Regarding sedimentation, since it works in a laminar regime, Stokes' Law can be applied to calculate the sedimentation velocity of the particles using Eq. 5.29.

$$v_s = \frac{2(\rho_p - \rho_f) \cdot g \cdot d_p^2}{9\mu_f} \tag{5.29}$$

Where v_s is the sedimentation velocity, ρ_p the density of the particles, ρ_f the density of the fluid, g the gravity, d_p the diameter of the particles and μ_f the dynamic viscosity of the fluid.

The best way to minimize this sedimentation velocity is using **similar densities of the fluid and the particles**, so the velocity of the particles will be more representative of the velocity of the fluid.

A rough evaluation of particle motion can be obtained using the Merzkirch Equation:

$$\frac{dU_p}{dt} = K(U_f - U_p) \tag{5.30}$$

In the previous expression, U_f and U_p are the velocities of the fluid and the particles, respectively. K is a constant that depends on the properties of the fluid and the particles.

Establishing as an initial condition that the particles are at rest, Eq. 5.30 is integrated and the behavior of the seeding particles against time is obtained.

$$U_p = U_f(1 - e^{-K \cdot t}) \tag{5.31}$$

It is usually called relaxation time, τ_s, to the time that elapses until the particle has reached 63% of the average velocity of the fluid, coinciding with the value of $1/K$. An expression for the relaxation time is collected in Vergini and Maddalena [18]:

$$\tau_s = \frac{4}{3} \cdot \frac{\rho_p d_p^2}{C_D Re_p \mu_f} \tag{5.32}$$

For the hypotheses of spherical particles and low Reynolds numbers referred to the particles, the drag coefficient $C_D = 24/Re_p$, being able to get the following expression:

$$\tau_s = \frac{\rho_p d_p^2}{18\mu_f} \tag{5.33}$$

2. **They must scatter enough light to be able to be registered**. This criterion not only depends on the seeding particles (refractive index with the surrounding medium and its polarization), but is also related to the power of the laser and the sensitivity of the recording system, and the observation angle.

   To study the scattering of light in spherical particles with a diameter d_p less than the wavelength of the laser incident light, λ, it can be used the scattering theory of Mie [17], characterized by the normalized diameter q:

$$q = \frac{\pi d_p}{\lambda} \tag{5.34}$$

When q is much less than 1, Rayleigh's theory is used, while when q is much greater than 1, geometric law is used. PIV technique uses Mie's theory because it is in an intermediate range where q is approximately unity. Generally, light scattered from 0° to 180° from a linearly polarized incident wave is linearly polarized in that same direction, and the scattering efficiency is independent of polarization, while the scattering efficiency at other observation angles is highly dependent of the polarization of incident light on particles [17].

In Fig. 5.12, it can be seen that the maximum local q appears in the angular distribution over the range of 0° to 180°. This supposes a disadvantage of the PIV technique because the usual configuration of the registration system at 90° with respect to the incident light of the laser, much of the light is scattered forward and a lot of intensity is lost, so it is necessary to use more powerful lasers.

3. **There must be the right amount of particles to get good results**. The last requirement is the result of experience and various simulations, but it is often used as a criterion that each interrogation area contains at least 8 particles [19].

For experimental development, Fluorescent Polymer Particles (FPP) with Rhodamine-B coating have been considered because they have spherical shape, density similar to that of the mixture of water and glycerin, not very small size and they

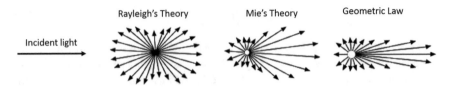

Fig. 5.12 Scattering for different particle sizes

Table 5.5 Characteristics of some seeding particles

Supplier	Dantec dynamics	Microparticles GmbH
Particles material	Fluorescent Polymer	Melanin resin
Average size	10 μm (1–20 μm)	9.84 μm
Density	1.19 g/cm^3	1.51 g/cm^3
Terminal velocity (agua + glicerina)	15 μ m/min (sedimentation)	164 μ m/min (sedimentation)
Terminal velocity (NaI solution)	999 μ m/min (floatation)	418 μ m/min (floatation)
Relaxation time (water + glycerin)	0.94 μs	1.2 μs
Relaxation time (NaI solution)	3.7 μs	4.7 μs

are phosphorescent. They must be phosphorescent so that the wavelength reflected by them when the laser sheet hits them is detected by the camera. The incident light from the laser is 532 532 nm, so the light emitted by the particles must be higher. Table 5.5 shows the main characteristics of the seeding particles proposed. Some of these parameters, such as sedimentation velocity or relaxation time, depend on the working fluid. In case of using NaI solution as fluid, the most appropriate particles would be made of melanin resin because the density is more similar.

If it is required a detailed study of a specific area of the model, smaller particles (5 μm) could be used using the microscope lens with the Micro-PIV technique.

5.3.3 Image Processing

The acquisition and processing of images using the PIV technique is usually carried out using specific software, such as the DynamicStudio package.

Once all the images have been captured and stored, they will be processed according to the following phases:

- **Calibration of the images**. It must be indicated the correspondence between the distances of the images taken and the real distances of the model. It is very important to do this process with precision because the error made in the calibration process can cause deviations of the values of the variables obtained with respect to reality. As a result, the magnification factor M will be obtained, which relates the real distance of two points (L_{real}) and the distance of those same points in the acquired image (L_{img}), as Eq. 5.35 shows.

$$M = \frac{L_{img}}{L_{real}} \tag{5.35}$$

Note that the magnification factor is the inverse of the scale factor, which is the one obtained with the DynamicStudio software.

- **Mask application**. In this step, the model limits (walls) are established where the non-slip condition is imposed. This is a process that is done manually, so you have to be careful and act as precisely as possible. It must be done every time the camera is moved since the position of the model changes.
- **Filtering images**. In this step, noise and defects in the images in the duct areas are eliminated. This is accomplished removing the background from the images and subtracting the image from minima.
- **Calculation of the velocity fields of each pair of images**. A cross-correlation is applied between each pair of images with the processing parameters that are established, mainly the interrogation area and overlapping.
- **Filtering of velocity fields**. Inconsistent velocity vectors are detected and replaced by vectors with a value equal to the average of those closest to their position.
- **Calculation of the average velocity field**. An ensemble averaging of the velocity fields obtained from each pair of images is used.
In case of pulsating flow, it refers to the average of the velocity fields at the same instant in the cardiac cycle over multiple consecutive cycles [9]. It is represented by Eq. 5.36.

$$\overline{U}(\mathbf{x}, t) = \frac{1}{N} \sum_{n=1}^{N} U^{(n)}(\mathbf{x}, t) \tag{5.36}$$

Where $U^{(n)}(\mathbf{x}, t)$ refers to the velocity vector in cycle n at the spatial location (x, y, z) at the measurement instant t of the cardiac cycle. $\overline{U}(\mathbf{x}, t)$ is the assembled average of the velocity vector and depends on the spatial location and the measurement instant of the cardiac cycle.

- **Other calculations**. Data obtained from the velocity fields can be exported to Tecplot code to perform post-processing. Thus other parameters can be obtained, such as the velocity gradient or the streamlines.

When using the PIV technique, certain parameters must be established, both while taking measurements and during their processing. The adjustment of these parameters with a suitable criterion is essential to obtain correct results.

Regarding measurements, the following parameters can be controlled:

- **Time between images**: elapsed time between each pair of captured images. This step is important because the program will determine the velocity profile of the fluid relating the time elapsed between two successive images and the displacement of each particle in the fluid field. If the time between images is too long it can lead to erroneous velocities since the particle may have moved outside the interrogation area. So if the flow varies in the model, the time between images must be adjusted. The commonly accepted criterion says that this time must be adjusted so that the particles move from one image to another no more than 10 pixels. [19]. The time between images can be calculated using Eq. 5.37.

$$t_{img} = \frac{n_{pixel} \cdot d_{pixel} \cdot r_{length}}{M \cdot v} \tag{5.37}$$

In the expression above, n_{pixel} is the number of pixels of the interrogation area in the axial direction, d_{pixel} is the size that a pixel occupies in the sensor, M is the magnification factor that relates the actual length to the image length (obtained in calibration), r_{length} refers to the relation in percentage between the length traveled by the particles and the square root of the interrogation area, and v is the fluid velocity in that section.

- **Shooting frequency**. This is what is called the sampling frequency. In steady flow, the maximum allowed value will be used.

  For pulsating flow, low-repetition rate PIV will be used to acquire data at frequencies much lower than the dominant frequencies of the flow. Thus, data is acquired at a given instant in the cycle over multiple cycles and then averaged per phase for analysis [9].

  An analysis can also be carried out at frequencies of the flow, in what is called spectrum. Mathematically, the spectral analysis is related to the Fourier transform, which relates a function in the time domain with a function in the frequency domain.

  For a signal in time $f(t)$ and a signal in the frequency domain $F(\omega)$, the Fourier Transform (Eq. 5.38) and the Inverse Fourier Transform are verified (Eq. 5.39)

$$F(\omega) = \int_{-\infty}^{\infty} f(t) \cdot e^{-i\omega t} \cdot dt \tag{5.38}$$

$$f(t) = \frac{1}{2\pi} \int_{-\infty}^{\infty} F(\omega) \cdot e^{+i\omega t} \cdot d\omega \tag{5.39}$$

Component frequencies are represented as peaks in the frequency domain. In this way, it is very useful to transform a signal into the frequency domain to obtain information that is not evident in the temporal domain. Figure 5.13 illustrates this explanation.

- **Number of images**. It is the number of pairs of images taken. Only one image is required to calibrate the image. For the rest of the measurements, depending on the uncertainty you want to have, more or less pairs of images will be chosen. This aspect is explained in more detail in the next section, but it can be said that about 100 pairs of images are sufficient.

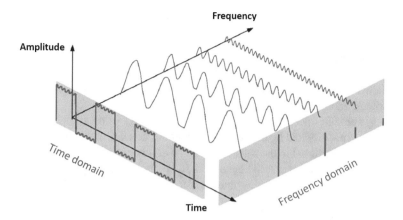

Fig. 5.13 Scheme of the Fourier Transform (Creative Commons License (CC0): public domain)

During the image processing, certain parameters must be established to obtain the velocity fields:

- **Processing type**. Basically, you can choose between two types of processing. One is based on calculating the cross-correlation of each pair of images, giving a velocity field for each one of them, and the other after doing this, averages all the velocity fields obtained, resulting in only one. Depending on the objective of each measurement, one or the other will be used. For the study of steady flow, averaging is shown as the best option, as in the case of pulsating flow, where each pair of images will be taken at the same instant of the cycle. On the other hand, for a frequency analysis of the velocity, it does not make sense to obtain the averaged velocity field, having to use the one provided by each pair of images.
- **Interrogation Area (IA) size**. This parameter has a direct influence on the level of detail of the velocity field. The IA is the region where the average displacement of the particles is calculated. If it is very large, the detail is lost when averaging over a large area. If it is excessively small, however, the point can be reached where there are not enough particles inside it so that the algorithm calculates the displacement, giving an erroneous result. The choice looks for a compromise solution. The commonly used criterion is that each IA must contain at least 8 particles to ensure a good result [19], and the size is decided based on this.
- **Overlapping**. This parameter allows to give a higher density to the velocity field, but without increasing its level of detail. What it does is overlap the IAs to create more points to calculate the velocity. It does not give more detail, so it does not modify the uncertainty of the result. It can be set between 25% and 50% to present a smoother velocity field.

5.3.4 Uncertainty Calculation

5.3.4.1 Velocity Uncertainty

Uncertainty in velocity measurement using a PIV system can be divided into two categories: systematic uncertainty and random uncertainty [9].

The systematic uncertainty (bias) constitutes a fixed value associated with the system and is not affected by the number of samples. Instead, the random uncertainty is inversely proportional to the square root of the number of samples [20]. To achieve statistical convergence, it is sufficient to take 100 pairs of images. With this number of images, the error in the mean value obtained for each vector is below 1%. The velocity uncertainty ($\Delta v/v$) is estimated according to the law of propagation of uncertainties from the uncertainty in the magnification factor ($\Delta M/M$), the time between images ($\Delta t_{img}/t_{img}$) and the calculation of the displacement of the particles in the images ($\Delta X_{pix}/X_{pix}$) as indicated in Eq. 5.40:

$$\left(\frac{\Delta v}{v}\right)^2 = \left(\frac{\Delta X_{pix}}{X_{pix}}\right)^2 + \left(\frac{\Delta M}{M}\right)^2 + \left(\frac{\Delta t_{img}}{t_{img}}\right)^2 \tag{5.40}$$

To know uncertainty associated with the velocity measurement using PIV, the uncertainties associated with the terms that appear in Eq. 5.40 must be estimated.

The uncertainty in the time between images is determined by the PIV system.

The error on the scale will depend on the camera configuration. Using Eq. 5.41, the uncertainty associated with the magnification factor is obtained.

$$\frac{\Delta M}{M} = \sqrt{\left(\frac{\Delta L_{img}}{L_{img}}\right)^2 + \left(\frac{\Delta L_{real}}{L_{real}}\right)^2} \tag{5.41}$$

The uncertainty in calculating the displacement of the particles in the images is estimated as a function of the peak correlation ratio, as proposed by Charonko et al. [21]. Using the height and width of the peaks of the cross-correlation, the quality of the vector map obtained can be evaluated. Ideally, the height of the first peak should be greater than that of the second peak, resulting in a peak ratio greater than unity. The processing method is usually the SCC standard (uses the Fast Fourier Transform FFT). So the uncertainty associated with the displacement of the particles in the measurements can be calculated with the peak ratio following Eq. 5.42.

$$\frac{\Delta X_{pix}}{X_{pix}} = \sqrt{\left(13.1 \cdot e^{-\frac{1}{2}\left(\frac{R_P-1}{0.327}\right)^2}\right)^2 + \left(0.226 R_p^{-1}\right)^2 + (0.08)^2} \tag{5.42}$$

In the previous expression, ($\Delta X_{pix}/X_{pix}$) is the displacement uncertainty and R_p is the ratio of the height of the peaks in the cross-correlation, $P1/P2$. According to the previous values, the **total uncertainty of the velocity measurement** is bounded lower by the value of 8%.

5.3.4.2 Reynolds Number Uncertainty

The Reynolds number is calculated using Eq. 5.43.

$$Re = \frac{\rho v D}{\mu} \qquad (5.43)$$

The uncertainty associated with the indirect calculation of the Reynolds number is obtained through the uncertainties of the density, velocity, and dynamic viscosity of the fluid, in addition to the partial derivatives of the Reynolds with respect to these variables obtaining Eq. 5.44:

$$\frac{\Delta Re}{Re} = \sqrt{\left|\frac{vD}{\mu}\right|^2 \cdot \left(\frac{\Delta \rho}{\rho}\right)^2 + \left|\frac{\rho D}{\mu}\right|^2 \cdot \left(\frac{\Delta v}{v}\right)^2 + \left|\frac{\rho v D}{\mu^2}\right|^2 \cdot \left(\frac{\Delta \mu}{\mu}\right)^2} \qquad (5.44)$$

Previously, it has been shown how to obtain the uncertainty in the velocity measurement ($\Delta v/v$).

Gravimetric methods are often used to determine the density of the fluid. The density ρ is calculated using Eq. 5.45, where m is the mass present in the volume that is entered in the precision scale:

$$\rho = \frac{m}{V} \qquad (5.45)$$

The error associated with the density measurement is calculated using Eq. 5.46:

$$\Delta \rho = \left|\frac{\partial \rho}{\partial m}\right| \cdot \Delta m + \left|\frac{\partial \rho}{\partial V}\right| \cdot \Delta V = \frac{1}{V} \cdot \Delta m + \frac{m}{V^2} \cdot \Delta V \qquad (5.46)$$

Δm is the error associated with the mass measurement and ΔV the error of the volume measurement.

The values of the density and especially of the dynamic viscosity vary with the temperature of the fluid. For this reason, the fluid temperature must be continuously monitored. The uncertainty in temperature ($\Delta T/T$) will depend on the measuring instrument.

The estimation of the uncertainty associated with the dynamic viscosity ($\Delta \mu/\mu$) could be obtained from the empirical expressions of the dynamic viscosities if the viscosity of the fluid is not measured directly or through the uncertainty associated with the fluid viscosity measurement equipment.

References

1. San Román JA, Fernández RS, García ER, Fernández-Avilés F (2006) Conocimientos básicos necesarios para realizar resonancia magnética en cardiología. Revista Española de Cardiologia 6(Supl.E):7–14
2. Hernández C, Zudaire B, Castaño S, Azcárate P, Villanueva A, Bastarrika G (2007) Principios básicos de resonancia magnética cardiovascular (RMC): secuencias, planos de adquisición y protocolo de estudio. Anales del Sistema Sanitario de Navarra 30:405–418
3. Gálvez M, Farías M, Asahi T, Bravo E (2005) Cálculo de tiempos t1 y t2 in vitro. Revista chilena de radiología 11:109–115
4. Hollnagel DI, Summers PE, Poulikakos D, Kollias SS (2009) Comparative velocity investigations in cerebral arteries and aneurysms: 3d phase-contrast MR angiography, laser doppler velocimetry and computational fluid dynamics. NMR in Biomed 22(8):795–808
5. Yan H, Liu Z (2001) The uncertainty relationship in magnetic resonance imaging (mri). arXiv Physics e-prints, September 2001
6. Xu H, Othman SF, Magin RL (2008) Monitoring tissue engineering using magnetic resonance imaging. J Biosci Bioeng 106(6):515–527
7. Ha H, Kim GB, Kweon J, Lee SJ, Kim Y-H, Lee DH, Yang DH, Kim N (2016) Hemodynamic measurement using four-dimensional phase-contrast MRI: quantification of hemodynamic parameters and clinical applications. Korean J Radiol 17(4):445
8. Dyverfeldt P, Bissell M, Barker AJ, Bolger AF, Carlhäll C-J, Ebbers T, Francios CJ, Frydrychowicz A, Geiger J, Giese D, Hope MD, Kilner PJ, Kozerke S, Myerson S, Neubauer S, Wieben O, Markl M (2015) 4d flow cardiovascular magnetic resonance consensus statement. J Cardiovascular Magn Reson 17(1)
9. Raghav V, Sastry S, Saikrishnan N (2018) Experimental assessment of flow fields associated with heart valve prostheses using particle image velocimetry (piv): Recommendations for best practices. Cardiovascular Eng Technol 9(3):273–287
10. Prasad AK (2000) Stereoscopic particle image velocimetry. Exp Fluids 29(2):103–116
11. Elsinga GE, Scarano F, Wieneke B, van Oudheusden BW (2006) Tomographic particle image velocimetry. Exp Fluids 41(6):933–947
12. Yazdi SG, Geoghegan PH, Docherty PD, Jermy M, Khanafer A (2018) A review of arterial phantom fabrication methods for flow measurement using piv techniques. Ann Biomed Eng 46(11):1697–1721
13. Wright SF, Zadrazil I, Markides CN (2017) A review of solid-fluid selection options for optical-based measurements in single-phase liquid, two-phase liquid-liquid and multiphase solid-liquid flows. Exp Fluids 58(9)
14. Cheng N-S (2008) Formula for the viscosity of a glycerol-water mixture. Ind Eng Chem Res 47(9):3285–3288
15. Bai K, Katz J (2014) On the refractive index of sodium iodide solutions for index matching in PIV. Exp Fluids 55(4)
16. Najjari MR, Hinke JA, Bulusu KV, Plesniak MW (2016) On the rheology of refractive-index-matched, non-newtonian blood-analog fluids for PIV experiments. Exp Fluids 57(6)
17. Raffel M, Willert CE, Scarano F, Köhler CJ, Wereley ST (2018) Particle image velocimetry. Springer GmbH
18. Vergine F, Maddalena L (2014) Stereoscopic particle image velocimetry measurements of supersonic, turbulent, and interacting streamwise vortices: challenges and application. Progr Aerosp Sci 66:1–16
19. Willert CE, Gharib M (1991) Digital particle image velocimetry. Experiments in Fluids 10(4):181–193
20. Coleman HW, Steele WG (2009) Experimentation, validation, and uncertainty analysis for engineers. Wiley
21. Charonko JJ, Vlachos PP (2013) Estimation of uncertainty bounds for individual particle image velocimetry measurements from cross-correlation peak ratio. Measur Sci Technol 24(6):065301

Correction to: Fluid Mechanics Applied to Medicine

Correction to:
A. Pozo Álvarez, *Fluid Mechanics Applied to Medicine,*
SpringerBriefs in Computational Mechanics,
https://doi.org/10.1007/978-3-030-60389-2

The initially published version of author's last name in Refs. 3, 4, 9 of Chap. 3, Refs. 22, 23, 30, 33 of Chap. 4 and Refs. 2, 3, 6, 8, 13, 15 of Chap. 5 were incorrect. The author's last names are corrected in References. The erratum chapters and the book have been updated with the change.

The updated versions of these chapters can be found at
https://doi.org/10.1007/978-3-030-60389-2_3
https://doi.org/10.1007/978-3-030-60389-2_4
https://doi.org/10.1007/978-3-030-60389-2_5